Starter Guide for
POS Device Applications Using .NET

By
Sean D. Liming and John R. Malin

Annabooks®

Copyright

Dedication

Dedicated to friends, family, and all those who worked on point of services/sales programs at Microsoft.

Table of Contents

Preface

Almost two decades as of this writing, I published the book *Windows Embedded for Point of Service / Pos for .NET Step-by-Step*. The book was the follow-on to the WEPOS and POS for .NET presentations at the Microsoft Embedded Developers Conferences (MEDC) in 2006. Since that time, there have been many changes in the POS industry and Windows programming. I followed up with two more books: *Professional's Guide to POS for .NET*, which covers POS for .NET 1.12 and *Professional's Guide to POS for Windows Runtime*, which covered POS for WinRT in UWP applications.

The POS books have been very popular and have sold all over the world. POS for .NET 1.14.1 was released in 2017, and there don't appear to be any updates on the horizon. POS for WinRT was limited to only supporting UWP applications, a few years ago, but Microsoft has moved quickly to improve WinUI and .NET Core over the past few years. With .NET 8, the POS for WinRT APIs are fully supported in Windows, so you can create WPF .NET 8 applications to interact with POS devices. The timing is right to refresh the POS for .NET and POS WinRT topics.

This book brings the two topics of POS for .NET and POS for WinRT together. The logical flow of the book is to cover POS for .NET first and then POS WinRT. The goal is to show how POS WinRT with .NET 8 provides a more modern approach to interacting with POS devices.

POS devices have a long lifespan and my POS equipment is still usable even after a couple of decades. It always is fun to break out the POS hardware to work on these books. Some happy memories came back as I worked on the exercises.

Like the previous books, the examples stick to the basic POS devices I have available to test with. I hope that you find this book to be a good foundation for your POS project. Again, please don't hesitate to provide feedback.

Sean D. Liming
Rancho Mirage, CA

Acknowledgments

This is the 4th book Annabooks has published that covers POS applications on Windows. There are several people we wish to thank for their past and present support for the first three books and this book. Our apologies if we forgot anyone.

At Microsoft, Mike Hall asked Sean to speak at MEDC India 2006 on WEPOS and POS for .NET. Mike was instrumental in making the contacts on the WEPOS / POS for .NET team. Craig Jensen provided invaluable technical assistance on the details of Service Objects, the POS for .NET architecture, and programming problems. He answered many of my initial questions and was able to get me updates for some of the undocumented features. David Baker was helpful in locating POS vendors and supporting the second book. Sylvester La Blanc took over for Craig and provided details during the years between the first and second books. Terry Warwick, Michael Chen, and Tom Kennard were instrumental in providing feedback and technical assistance for the third book.

Several of the POS device manufacturers have been very helpful over the years. Cary Eckmann of Hand-Held Products, now Honeywell, was very helpful in the use of the IT5600 scanner for the original book. Kiran Gandhi of MagTek was willing to share details on the MagnePrint technology his company was developing. The technology is incorporated in their new Dynamag secure card reader. Special thanks go to Douglas Evans and Bobby Billingsley of Wincor-Nixdorf. They were kind enough to donate hardware that was used in the last book.

There are two posters from the old MSDN forum with the handles of YORT and Woodchux that were active in their support for POS for .NET.

Finally, we thank TabletKiosk® for providing us with a TufTab i60XT for the third book. The mobile device lived up to its name in performance and durability.

Annabooks

Annabooks provides a unique approach to embedded/IoT system services with multiple support levels. Our different offerings include books, articles, training, and project consulting. Current publications and courses have focused on embedded PC architecture and Windows Embedded, which reach a wide audience from Fortune 500 companies to small organizations. We will continue to expand our future services into new technologies and unique topics as they become available.

Books

Starter Guide for STM32Cube™ and Eclipse ThreadX®

Starter Guide to Windows 10 IoT Enterprise

Java and Eclipse for Computer Science

Open Software Stack for the Intel® Atom™ Processor

Real-Time Development from Theory to Practice

The PC Handbook

Training Courses

Windows® 10/11 IoT Enterprise Training Course

Please contact us for more information on consultation and availability.

Annabooks®

Web: www.annabooks.com

1 Introduction

Point of Sale (POS) devices have gone beyond the traditional retail cash register into other market segments, such as shipping/receiving, inventory control, agriculture, banking, security systems, vending machines, gaming systems, and travel just to name a few. What contributed to the growth and expansion of the POS market into so many different market segments were industry standards and programming solutions that made writing POS applications easier. Microsoft Windows has been a driving force as the leading operating system in the retail space. In addition, Microsoft has created API SDKs and retail application solutions. This book will focus on the POS APIs and built-in support in Windows to write .NET and .NET Framework applications that interact with POS devices. Before we dive in, this introductory chapter provides a little historical background and the software and hardware items needed to write applications.

1.1 A Little History

OLE for POS (OPOS) was first introduced in 1996 as an implementation of the UnifiedPOS standard for Microsoft operating systems. The basic idea was to put an abstraction layer between the POS applications and the POS hardware. Developers would then be able to write one application that would support multiple POS hardware devices. If a POS hardware device had to be replaced, a simple modification with a control utility would allow the unmodified POS application to use the new hardware. OPOS was initiated by Microsoft, NCR, Epson, and Fujitsu-ICL. The UnifiedPOS specification was originally created by the Association for Retail Technology Standards (ARTS), and it is now governed by the Object Management Group®.

As Microsoft's operating systems evolved, so did the languages that programmers used to write applications. Microsoft introduced .NET Framework in 2000, and C# and VB.NET quickly became the programming languages of choice in which to write applications. In 2006, Microsoft introduced POS for .NET as a new .NET implementation of the UnifiedPOS specification. POS developers could write applications in C# and VB.NET to access devices via OPOS drivers or the new POS for .NET Service Objects (SOs). At the time, Microsoft referred to POS as Point of Service rather than Point of Sale. The concept was that POS devices can be used in different applications and not just a cash register. The change in acronym didn't catch on in industry, but Microsoft still calls this out in their documentation.

A lot happened in a few short years after the release of POS for .NET. Mobile devices became the next wave of computing. Apple introduced the iPhone and Google introduced Android both of which offered a simpler interface to Microsoft's Windows Phones. Both companies developed online stores to download software directly to mobile devices. Microsoft was caught a little flat-footed, but they moved to catch up. Microsoft introduced the Windows Presentation Foundation (WPF) to separate the GUI interface from the code. WPF allowed developers to simply modify the GUI for the targeted device using a simple extensible markup language. The Windows Operating system story had to evolve as well. Windows NT used to support different processors like MIPS, DEC Alpha, and Motorola, but those ports of Windows were dropped over time because Intel Architecture dominated the market. The ARM processor technology started to dominate the mobile landscape in the mid-2000s, and an effort was made to make Windows NT run on ARM. Eventually, Windows CE and Windows Embedded operating systems were end-of-lived, and Windows evolved into Windows 10, one Windows operating system that can reach platforms big and small. The only thing missing was an online store.

By the time Microsoft got around to developing their online stores, consumers were already comfortable buying anything on the Internet. The days of going to a store and buying software in a box were long behind us. The simple click and purchase from an online marketplace was the new normal. The key to success would be security. A new programming paradigm was needed. Starting with Windows 8, Microsoft introduced the concept of Modern Apps. Modern Apps evolved into Universal Windows Platform (UWP) programs. UWP programs supported different processor architectures and allowed programmers to reach a wide range of Windows devices. The new programming model came with limitations. .NET Framework applications and most prior Microsoft operating system applications had a certain degree of freedom in what could be accessed. The only limitation was accessing hardware directly. UWP applications live in a sandbox. Developers have to enable certain resources and users have to accept that applications can access these resources on their device. Since software could quickly be downloaded and installed, certain safety measures had to be put in place to make sure rogue applications wouldn't compromise a system. It was these limitations that made UWP applications very unpopular. During this time, Microsoft developed POS for Windows Runtime so developers can access a few of the popular POS devices via UWP applications.

The death of Windows Phone quickened the death of UWP. Out of the ashes rose .NET Core, which later got shortened to .NET. .NET Framework still exists and will be available for some time, but .NET Core is now the new programming paradigm that resolved the shortcomings of UWP, supported mobile devices, and applications that can run on ARM processors. Best of all .NET Core supports accessing the POS for Windows Runtime namespace so you can write POS applications in .NET Core.

As a result of all the history and development. Developers can create POS applications in either .NET Framework or .NET Core. This book explores both development paths.

1.2   What is POS for .NET Framework and POS WinRT: A Brief History

A single POS system can have multiple POS devices attached, such as magnetic card readers, scanners, cash drawers, and line displays. There are many POS device manufacturers in the market, and each manufacturer has a unique protocol for their device. Starting in 1995, the various POS vendors and manufacturers came together to create the OLE for POS (OPOS) specification. The specification addressed the common problem of not having a uniform programming interface when writing POS applications for devices from different manufacturers running on Microsoft Windows 3.x. The OPOS specification creates an abstraction layer between the application and the hardware device drivers. Application developers only need to write their applications using OLE or ActiveX objects to interact with devices from different manufacturers. The application calls a logic device name to interact with the hardware. JavaPOS soon followed in 1997 for Java developers.

Around 1998, the Association of Retail Technology Standards (ARTS) merged both OPOS and JavaPOS specifications to create the Unified POS standard (UnifiedPOS). In 2017, the National Retail Foundation merged with the Object Management Group® (OMG) which now manages the UnifiedPOS specification. UnifiedPOS defines a common API interface for different POS hardware devices.

POS for .NET is a .NET Framework implementation of the UnifiedPOS standard, and it takes advantage of the Plug and Play capability found in the Windows operating system. The basic concept is to abstract the application from the POS hardware. As hardware changes, all one has to do is replace the hardware and install a driver, and the application never has to be recompiled. POS for .NET also introduces a new Service Object model from OPOS, but it can still talk to OPOS drivers. POS for .NET SDK supports all 36 devices in the UnifiedPOS standard.

POS Windows Runtime (WinRT) doesn't require a separate SDK. All the support for POS devices is built into Windows starting with Windows 10 Build 1607 version 14393. The Windows.Devices.PointOfService namespace is part of the Universal Windows assembly. A developer has everything needed in Visual Studio to create a POS WinRT application, but support is limited to only 5 of the 36 POS devices: barcode scanners, cash drawers, line display, magnetic stripe reader (MSR), and receipt printers. The remaining 31 devices can be accessed as services via named pipes.

1.3   About the Book

There are two programming paradigms for writing POS applications in Windows. The goal of the book is to present both solutions, so you can determine which is the best solution for your application. This book is a combination of two previous books: *Professional's Guide to POS for .NET* and *Professional's Guide to POS for Windows Runtime*. The information has been updated for use with Visual Studio 2022 and .NET 8.

The book is intended to be read from cover to cover since some topics in POS for .NET are important for POS WinRT. The next four chapters focus on POS for .NET. The remaining chapters are on POS WinRT with the final chapter bringing everything together.

Chapter 2 – Discusses the basics of the UnifiedPOS specification, provides an introduction to the POS for .NET architecture, the role of the Service Object and OPOS drivers, introduces and installs the SDK, and how to get started in picking POS devices.

Chapter 3 – Covers different example POS applications to access POS devices: barcode, MSR, Line display, receipt printer, and cash drawer.

Chapter 4 - Setup and field management of a POS device is an important part of the product lifecycle. This chapter will look at the different tools and features for managing POS Service Objects and OPOS drivers.

Chapter 5 – Microsoft developed the Service Object to improve upon OPOS drivers. Creating a Service Object for a POS device will be the focus of the chapter.

Chapter 6 – Introduces POS WinRT and how devices are enumerated. Several exercises use the barcode scanner as an example.

Chapter 7 – Discusses different example POS WinRT applications that access POS devices: MSR, Line display, receipt printer, and cash drawer.

Chapter 8 – Demonstrates how a POS WinRT application can access one of the remaining 31 devices using a server and named pipes.

Chapter 9 – The final chapter presents example applications that access multiple POS devices.

1.4   Honorable Mention Retail Management System (RMS)

RMS is part of the Microsoft Dynamics® suite of products. Dynamics addresses customer relations software designed to work with backend connectivity for today's Cloud-connected systems. At least that is my high-level definition of the Dynamics product line. RMS is the retail point of sale application that runs in the cash register. RMS implements UnifiedPOS to connect to local POS devices. A detailed discussion of RMS is beyond the scope of this book, but you can find more information on the Microsoft Dynamics website.

1.5   Software / Hardware Requirements for the Exercises

In keeping with the "learn by doing" theme, there are multiple hands-on exercises throughout the book. To run these various exercises, you will need specific software and hardware to accomplish that goal. When it comes to hardware, you may make some substitutions. You might not be able to find the exact models of devices used in the book examples as some products may go end-of-life. Chapter 2 discusses how to look for POS devices that will work with POS for .NET.

Every effort has been made to ensure the accuracy of the examples and contents in the book. Please be aware that these are example applications and not production code. From my past book experience, I have learned that mistakes can sometimes slip through our editing and review cycles; so please check the website (www.annabooks.com) for updates.

1.5.1   Development System
The following are the hardware requirements for the development system:

- Windows 10 2H1 or Windows 11 or higher
- Intel-compatible processor 1.6GHz or higher
- 4GB or higher RAM
- 40GB or higher hard drive
- 10/100/1000 Ethernet
- Keyboard, mouse, display, etc.
- USB, serial, and parallel ports to interface to POS peripherals

1.5.2   POS Hardware
The following is the POS hardware used to develop the book exercises. Some of these devices are no longer available. Of course, you may substitute your hardware, but you may need to adjust for any differences in the setup.

- TabletKiosk TufTab i60XT or EMi61J Rugged Tablet with built-in barcode scanner (Honeywell N3680) and Desktop Dock.
- Honeywell (HHP) Barcode Scanner – IT5600 (5600SR050) USB/HID.
- MagTek USB/HID MSR – 21040101 or 21040102.
- Avery-Berkel (Weigh-Tronix) POS Scale 6710.
- EPSON T-T88V Receipt Printer with UB-E04 Ethernet interface.
- MMF Cash Drawer – VAL-u Line® MMFVAL 1313E04.
- EPSON DM-D110 Pole Display.

You can find a list of supported POS WinRT devices on Microsoft Learn: https://learn.microsoft.com/en-us/windows/uwp/devices-sensors/pos-device-support

1.5.3   Software
Below is a list of software that will be used in the exercises. Various exercises will list any extra software programs or drivers that may be required to complete the exercise.

- Windows 11 Professional / Ultimate or higher
- POS for .NET 1.14.1 or higher – check the Microsoft website for the latest releases
- .NET 8 runtime
- Visual Studio 2022 or higher
- SOManager from Annabooks (https://www.annabooks.com)

- Service Objects and OPOS drivers for the different POS hardware listed above. Please contact the POS manufacturer for more information.

1.6   Download Book Exercises

The original POS for .NET book covered both VB.NET and C#. C# has more support available, so VB.NET is not covered. All the projects are written in C#. The project examples can be downloaded from the book page at https://www.annabooks.com

1.7   Summary – Get Going with POS Development

POS systems have come a long way since the days of MS-DOS. UnifiedPOS presents the possibility for developers to focus on one application that can support a device from many different POS manufacturers. The best way to get going is to dive into some examples and learn by doing, but first, we need to install the POS for .NET SDK, which is covered in the next chapter.

2 POS for .NET Architecture and the SDK

Very simply, the idea behind .NET was to improve the overall application development process by using classes and namespaces. Maybe that was a little too simple of an explanation. My take on .NET is that managed code takes care of much of the memory management work and custom GUI needs, and it lets you concentrate on the application itself. POS for .NET is an implementation of the UnifiedPOS specification for C# and VB .NET developers, which allows developers to take advantage of data and device events. This chapter will cover the following:

- Core items from the UnifiedPOS specification
- POS for .NET architecture tied to the UnifiedPOS specification
- The importance of Service Objects and OPOS drivers
- Installation and review of the POS for .NET SDK
- Set up a barcode scanner and an MSR to use the SDK's example service objects

2.1 OLE for POS (OPOS) Architecture

Chapter 1 discussed the history behind the different OPOS, JavaPOS, and the new UnifiedPOS specifications. The goal of these POS specifications is to abstract the applications from the POS hardware so one application can support multiple POS devices. The architecture derived from these specifications provided the layer of abstraction. The OPOS specification had a Control Object and a Service Object sitting between the application and the hardware. Together, these two objects make up the OPOS driver or OPOS control. The illustration below shows the OPOS Architecture.

```
                    ┌─────────────────────┐
                    │     Application      │
                    └─────────────────────┘
                              ◇
              ┌───────────────┼───────────────┐
              │     ┌─────────────────────┐   │
              │     │    Control Object    │   │
              │     └─────────────────────┘   │       OPOS
              │               ◇               │       Driver
              │     ┌─────────────────────┐   │
              │     │    Service Object    │   │
              │     └─────────────────────┘   │
              └───────────────┼───────────────┘
                              ◇
                    ┌─────────────────────┐
                    │      Hardware        │
                    └─────────────────────┘
```

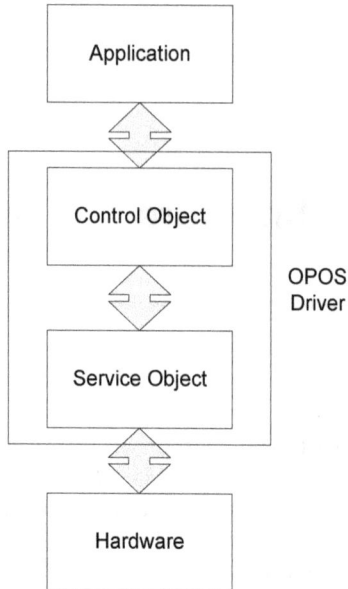

With OPOS, applications talk through ActiveX (OLE) or COM controls to the Control Object. There is only one ActiveX or COM control or Control Object per device class (cash drawer, scanner, coin dispenser, etc.). An application only needs to be coded once to access a single POS device, which can come from many different manufacturers. In the end, all OPOS driver developers use the same Control Object for their device class.

The differentiation lies in the Service Object. The Service Object contains the unique hardware-dependent functions that talk to the POS device. In the Service Object, the OPOS driver developer adds the unique code for their device. Communication between the Control Object and the Service Object is made through a combination of methods, events, and registry setup information.

To set up the registry information, a special setup application is needed. Administrators use an OPOS setup application to set up each device type for use with the application. POS application developers only need to provide the information on how to set up each POS device.

2.2   UnifiedPOS Summarized

POS for .NET SDK 1.14.1 follows the UnifiedPOS V1.14.1 specification. Before we dive into the POS for .NET architecture, it is important to cover some of the basics of the UnifiedPOS specification. The specification is over 2000 pages in length, so it is not possible to go into all of the details. A summary of key features will be provided. The specification defines the standard methods and properties for applications and Service Objects.

2.2.1　Applications

For applications, there are some basic methods and properties that are defined. The most commonly used methods are for accessing a device for usage:

1. OPEN – Open access to the Service Object or OPOS driver
2. CLAIM – Get exclusive access to the device
3. DeviceEnable – Enable the device
4. Perform device action
5. DeviceDisable – Disable the device
6. RELEASE – Release exclusive access to the device
7. CLOSE – Close access to the device

This sequence of steps will be used for each device that an application will access. In some cases, the device might never be released or closed. Other devices might require the full sequence if they share a port with another device.

2.2.2　Service Object / OPOS driver

The largest amount of information in the specification is common to all POS devices. The specification calls out the properties and methods for 36 different POS devices. It also defines a general model for each device. There are diagrams showing the call sequence for each device, as well as, a state diagram.

Each device has its own unique set of actions to perform: bill dispenser dispenses bills; cash drawers open and close; receipts are printed; scales record weight, etc. The actions are broken out and defined in different methods. The specification also calls out the device capabilities (Cap) that can be queried.

The common methods for each device are to check the health of the device and allow manufacturers to add unique device functions via a DirectIO method.

Chapter 5 will demonstrate a custom Service Object where you can see the specification applied to a real solution.

2.3　POS for .NET Architecture

The challenge with OLE for POS was that it required a good understanding of ActiveX / COM programming. POS for .NET was developed to simplify the application and Service Object development without requiring a PhD in COM development. The POS for .NET architecture makes changes so that application vendors can write apps in the familiar .NET style, and Service Object developers can focus on their unique device's capability.

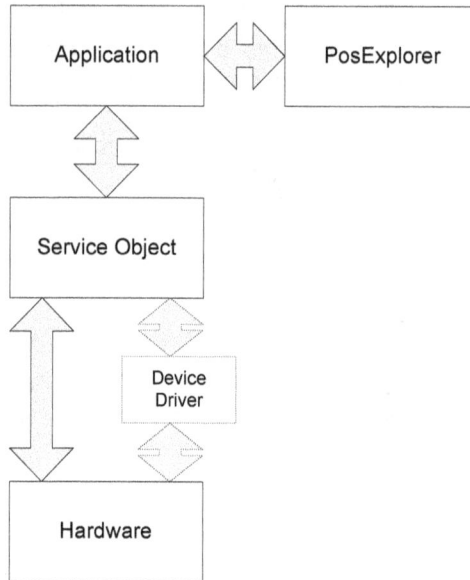

The biggest change is the removal of the Control Object, which is replaced by the PosExplorer to act as a class manager or class factory for all POS devices. Using the POS classes to create instances of the installed hardware, the application can interact with the Service Object. Although it was not shown in the OPOS diagram, the Service Object could interact with the device driver. Best of all, the new POS for .NET architecture still supports the old OPOS drivers using and interoperation through the old Control Object.

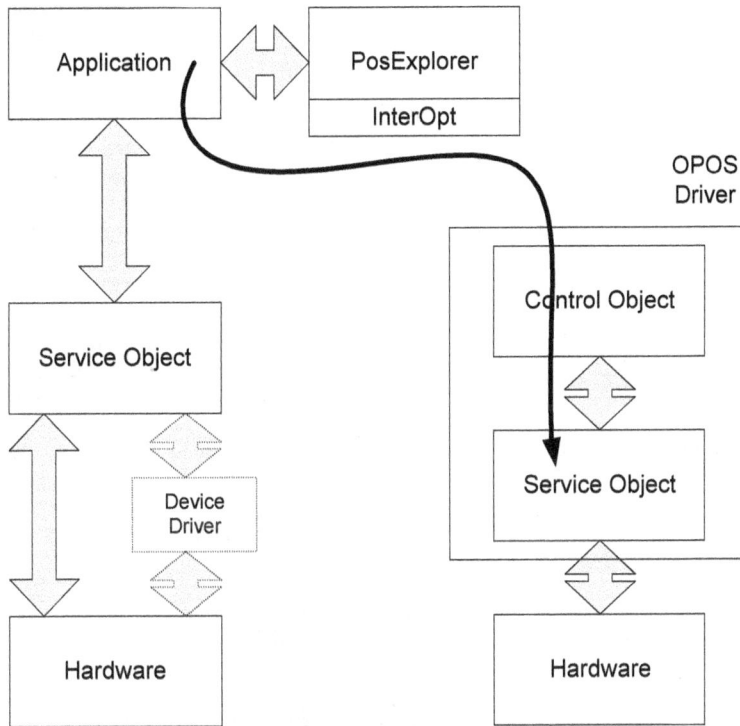

POS for .NET provides legacy COM interoperability support for a variety of devices, but not all POS devices are supported. Please see the POS for .NET SDK documentation for the up-to-date list of devices supported.

In place of ActiveX / COM calls, POS for .NET makes the link between the application and Service Object with a combination of registry keys, XML files, and standard directory structures. When an application starts, the registry is queried for the locations of the XML configuration files, Service Object assemblies, and legacy OPOS drivers. As the application adds instances of devices, the information in the registry and XML files is used to link the applications to the appropriate device Service Object. The table shows the core registry locations used by POS for .NET:

Registry Path (Windows 32Bit)	Registry Path (Windows 64Bit)	Description
HKLM\SOFTWARE\POSfor.NET \ControlAssemblies	HKLM\SOFTWARE\Wow6432Node \POSfor.NET\ ControlAssemblies	Defines the paths to the Service Object DLL files. Using this registry key, Service Object developers can define the location of their Service Object

		anywhere in the file system. Companies can use the standard install path for their software: \Program Files\<company name>
HKLM\SOFTWARE\POSfor.NET \ControlConfigs	HKLM\SOFTWARE\Wow6432Node \POSfor.NET\ControlConfigs	Configuration XML files are used with PnP Service Objects to link the HID hardware ID to the actual Service Object DLL, thus the XML file allows the Service Object to support multiple devices. The registry key holds the paths to the XML files.
HKLM\SOFTWARE\POSfor.NET\Setup	HKLM\SOFTWARE\Wow6432Node \POSfor.NET\Logging	POS for .NET logging information
HKLM\SOFTWARE\POSfor.NET\Setup	HKLM\SOFTWARE\Wow6432Node \POSfor.NET\Setup	POS for .NET setup information
HKLM\SOFTWARE\OLEforRetail \ServiceOPOS (\<type>\<device>)	HKLM\SOFTWARE\Wow6432Node \OLEforRetail\ ServiceOPOS (\<type>\<device>)	Used for legacy OPOS driver – logical name and setup information

Important Registry Location for POS Explorer

Keep in mind that POS for .NET is 32-bit and Windows 11 is only 64-Bit. The middle column is the location of these registry keys.

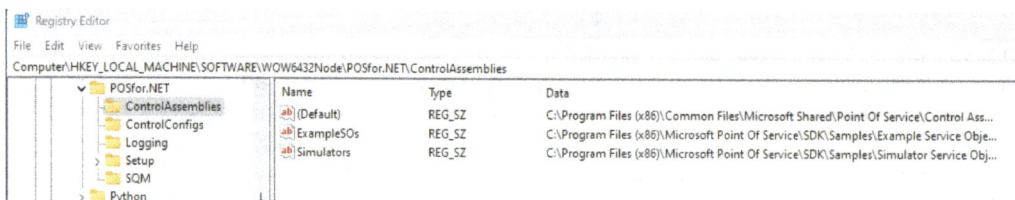

With POS for .NET, Service Object developers have a little more freedom to locate and support PnP devices:

- With the latest PnP devices and an XML file, Service Object developers can create one Service Object that supports many devices. An XML file is used to link the HID Hardware ID to the actual Service Object.
- With the ability to set the path to the Service Object in the registry, companies can perform a normal installation of their drivers to \Program Files\<company name>

OPOS drivers are still supported as PosExplorer uses the HKLM\ SOFTWARE\ OLEforRetail\ ServiceOPOS (For Windows 64-bit, the key is HKLM\ SOFTWARE\Wow6432Node\ OLEforRetail\ ServiceOPOS) registry key to track down OPOS drivers. Each device type has its own subkey under this registry key. Further, if there are a couple of scanners installed, each scanner will get its own subkey with the setup information under the scanner subkey. Some manufacturers may have their own custom OPOS driver implementation.

POS for .NET will also set up some standard files and directory paths.

- *C:\ProgramData\Microsoft\Point Of Service\Configuration\Configuration.xml* - The Configuration.xml file contains any custom setup Service Object information, and is typically used for non-PnP Service Objects. Not defined by the registry keys, a POS for .NET application will always look for the Configuration.xml file at this location.

- *C:\ProgramData\Microsoft\Point Of Service\Statistics\ StatisticsPosDeviceStatistics.xml* – If enabled in the Service Object, this XML file will contain Service Object information.

- C:\Program Files (x86)\Common Files\Microsoft Shared\Point Of Service\Control Assemblies – The standard location for Service Objects. You may choose to place your Service Objects in this standard directory or not. The path is pointed to by the HKLM\SOFTWARE\POSfor.NET\ControlAssemblies registry key.

- C:\Program Files (x86)\Common Files\Microsoft Shared\Point Of Service\Control Configurations – The standard locations for configuration XML files. You may choose to place your configuration XML file in this standard directory or not. The path is pointed to by the HKLM\SOFTWARE\POSfor.NET\ControlConfigs registry key.

These registry paths, XML files, and directories are standard when the POS for .NET SDK is installed.

2.4   POS for .NET Classes and Interactions

POS for .NET is a .NET implementation of the UnifiedPOS specification. POS for .NET keeps the application separate from the POS hardware through interactions with POSExplorer and the Service Object.

POS for .NET removes the Control Object defined by OPOS and replaces it with PosExplorer. PosExplorer acts as a class manager or class factory. This eliminates the Control Object / Service Object mismatches that have been found in UnifiedPOS, and it allows you to take advantage of Plug and Play events. Using the POS classes to create instances of the installed hardware, the application can interact with the Service Object. Best of all, the new POS for .NET architecture still supports the old OPOS drivers using interoperation through the old Control Object.

There are 6 core classes that come with POS for .NET:

- Interface Classes - POS for .NET Interface Base Classes for POS devices. Basic and Base Classes are derived from the Interface Classes.
- PosExplorer Class - Provides POS applications with a single-entry point to POS for .NET services.
- DeviceInfo Class - Supplies POS applications with information about POS devices and the Service Objects associated with them.
- Exception Classes - Error handling.
- Specific Interface Classes - Lists the specific Interface Classes that POS for .NET has pre-defined for programmers to use when implementing Service Objects.
- PosCommon Class - Describes the PosCommon class, the Base Class upon which each Service Object class or interface is built.

The main classes used to write applications are the PosExplorer, DeviceInfo, and Exception classes. Service Object developers will use derivatives of the Interface Class.

2.4.1 PosExplorer

The PosExplorer Class serves as the interface between Windows PnP notifications and POS applications. PosExplorer translates the relevant PnP notifications into POS for .NET events, which it then sends to the POS application.

Plug and Play Events - The PosExplorer Class exposes two PnP events for use by POS applications:

- DeviceAddedEvent triggers when a POS PnP device is connected to the system.
- DeviceRemovedEvent triggers when a POS PnP device is disconnected from the system

PosExplorer contains all the information about the POS devices in the systems. When a POS for .NET application first launches, PosExplorer looks for any Service Objects or OPOS drivers in the system by following the path of three different registry keys. Looking up registry keys allows OEMs to place Service Objects anywhere in the system but most likely in \program files\company name, where the Service Object can easily be upgraded.

- Service Object binary location paths are found in the:
  HKLM\SOFTWARE\WOW6432Node\POSfor.NET\ControlAssemblies

- XML files can be used for devices that share the same Service Object. The XML files can be placed in a path defined by the HKLM\Software\WOW6432Node\POSfor.NET\ControlConfigs registry key.

- PosExplorer can also locate the legacy OPOS drivers: HKLM\SOFTWARE\ WOW6432Node \OLEforRetail\ServiceOPOS (\<type>\<device>).

2.4.2 Application Interaction
A POS for .NET application interacts with the Service Object and PosExplorer through the use of events and invocations.

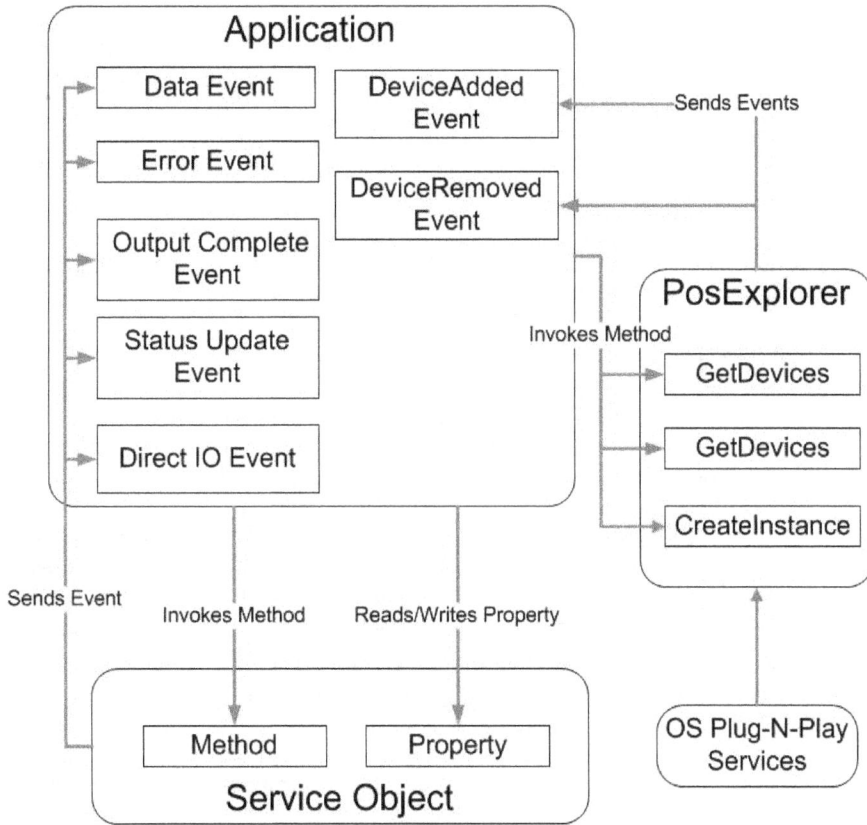

Application and Service Object Interaction - The Service Object can send 5 different events to the application: Data event, Error event, Output Complete event, Status event, and Direct IO event. The application can invoke different methods and check or change properties in a Service Object. The actual methods, properties, and events available differ from device to device.

Application and PosExplorer Interaction – PosExplorer monitors Plug and Play messages to determine if POS devices have been added or removed. A device add / remove event can then be sent to the application. The application can query information about available devices and obtain instances of devices to interact with the Service Object.

2.5 *Choosing POS Devices Carefully*
As you can see, the Service Objects and OPOS drivers play an important role in the device selection and setup of the system. Choosing a device without either driver is a typical

15

mistake developers new to POS for .NET make. Here is a basic guide to help with your development:

2.5.1 Choose POS devices that have Service Object or OPOS support

A POS device with a Service Object should be the first choice, since Service Objects are designed for POS for .NET. It has taken some time for the POS industry to develop Service Objects, and not all manufacturers have implemented Service Objects for their devices. If a Service Object is not available, an OPOS driver is the next best choice though not all OPOS drivers will work with POS for .NET. You may need to install the ActiveX Control Object to get the device to work with POS for .NET. MCS has been a popular place to get the common Control Objects: http://monroecs.com.

TIP: Always contact the POS device manufacturer to make sure they have either a Service Object or OPOS driver available for use with POS for .NET.

2.5.2 USB HID vs. Keyboard Wedge

Barcode scanners and MSRs come with a variety of interfaces – serial, USB, and custom. You will want to look for devices that have USB HID support to take advantage of Plug and Play support. The POS for .NET SDK's sample service-object POS supports USB HID devices, and we will be taking advantage of this in the upcoming projects.

You will find barcode scanners and MSRs that support a keyboard wedge interface. This means that the device acts like a keyboard to the system. POS for .NET doesn't handle keyboard input as a data input event. A keyboard wedge device could still be used with the application, but the user must first manually focus on the text control in the application before scanning a barcode or swiping a magnetic strip card. It is better to choose devices where POS for .NET can handle the data input event to make the application easy to use and more secure.

Barcode scanners are sophisticated devices, where some can be re-programmed to change the interface mode. Re-programming is as simple as scanning a few barcodes from a user manual. Out of the box, a barcode scanner is configured as a keyboard wedge. You will want to change the barcode scanner to be a pure USB device for use with POS for .NET. This means that as you deploy systems out to the various stores, you will have to make sure that the installers perform this change, or order the devices preconfigured from the factory.

2.5.3 Serial Interface and Create a Service Object

If there is no Service Object or OPOS driver available and the POS device is a simple device like a cash drawer or weight scale, creating a custom Service Object is possible. Chapter 6 covers creating a custom Service Object.

USB has become a dominant interface on PC platforms, but the serial interface is still used in many POS devices and systems. Pole Displays, weight scales, cash drawers, and receipt printers are just a few devices that come with a serial interface. The industry is moving slowly to USB as the PC platform continues to evolve. A serial interface is the easiest to program, so you would want to choose a device that has a serial interface if you have to develop a custom Service Object.

The device has to come with a user manual that discusses the communication protocol, so you can develop the Service Object. The device also has to be simple. Cash drawers and weight scales are simple devices that can be easily programmed. Receipt printers are very complex, and you should leave this to the manufacturer to create a Service Object. Creating a Service Object for a receipt printer has popped up in the forums over the past few years, and the advice is still the same: "Look for a receipt printer with a Service Object". The most popular receipt printers have POS for .NET support.

2.5.4   Create the Application to call the Default POS Device

This last point is getting a little ahead of the book, but it is something to introduce, now, and keep in the back of your mind as you go forward. To make the application as generic as possible, the application should be designed to open the default POS device and stay away from logical names as much as possible. Since you may use devices with OPOS drivers, avoiding a logical name might be impossible. If the application calls out a logical name, you will have to document or inform the installer / owner that the POS devices must be configured with a logical name.

Default vs. logical name might be a little puzzling, and we will get to this in the next chapter. First, we need to install and look at what is in the POS for .NET SDK.

2.6   Exercise 2.1 Install and Review the POS for .NET SDK

In this exercise, we will install POS for .NET SDK in Windows desktop and review the files and directories that come with the SDK. The standard registry keys, XML files, and directories that were discussed earlier will be created during installation. In addition, there are some sample Service Objects with source code, a sample application with source, a device management utility, and the .NET libraries for developing applications and Service Objects.

Software Requirements:

- POS for .NET SDK 1.14.1 – downloaded from Microsoft.

1. Download the POS for .NET SDK installer MSI from the Microsoft website.
2. The older SDKs came as a zip file with a setup.exe program. The latest installer is just an MSI. Once the SDK is downloaded to the local drive, run the installer.
3. When the install wizard starts, click Next to continue the installation.
4. Accept the Eula and click Next.
5. Select Yes or No on the next page for feedback, and click Next.
6. Keep the installation path the same, and click Next.
7. Installation options appear. By default, only the runtime will be installed. This means you can run the installer with the silent install command switches and only the runtime will be installed, which is important for deployments. The SDK is needed for development. Click on the SDK drop-down and select "Will be installed on local hardware drive".

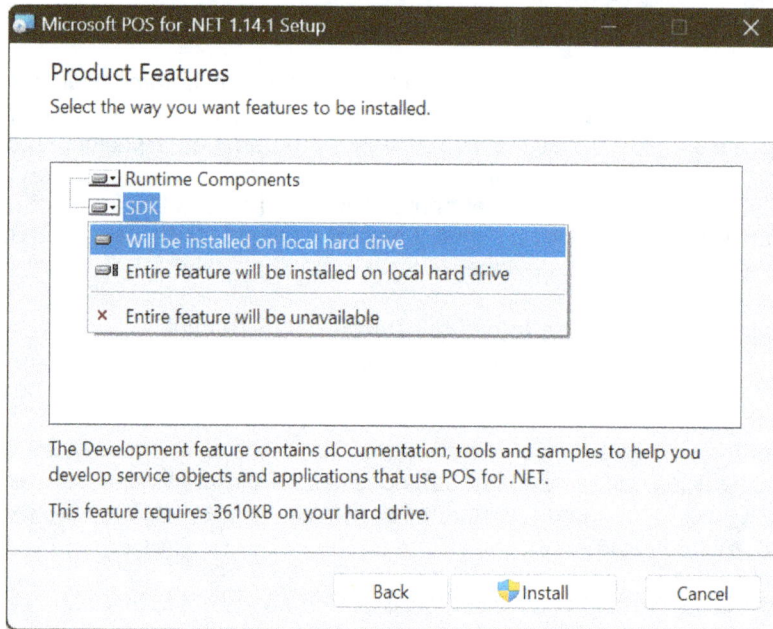

8. Click Install.
9. Click Finish after the installation has been completed.

Note: *POS for .NET is a 32-bit runtime. It will install the files in the \program files(x86) directory on a Windows 64-bit machine.*

Warning: POS for .NET is a little older SDK. Windows updates have been known to disrupt the SDK WMI API calls. You may have to re-install the SDK after Windows Update.

10. Once installed, open the regedit utility and you can see the installed registry keys discussed in section 2.3.
11. Once installed, open Explorer to review the directories discussed in section 2.3. In addition, the following directories and applications were added:

Directory	Description
C:\Program Files\Microsoft Point Of Service	Contains the POS Device Manager (POSDM.EXE) application and the POS for .NET runtime service. POSDM.EXE will be used to manage and set up POS devices, either locally or remotely.
C:\Program Files\Microsoft Point Of Service\1033	Contains the POS WMI Admin Message DLL.
C:\Program Files\Microsoft Point Of Service\SDK	The root of the SDK contains two important .NET libraries that will be used to develop applications and Service Objects – Microsoft.PointOfService.dll and Microsoft.PointOfService.CotrolBase.dll.

C:\Program Files\Microsoft Point Of Service\SDK\Documentation	contains the latest documentation
C:\Program Files\Microsoft Point Of Service\SDK\Samples\Example Service Objects	Binary and source code for a generic Service Object that supports barcode scanners and MSRs. The Service Object is written in C#.
C:\Program Files\Microsoft Point Of Service\SDK\Samples\Sample Application	Sample POS application and source code that can work with the simulator Service Object and real devices. The source code provides a good example of how to access devices using the POS for .NET classes. The application is written in C#.
C:\Program Files\Microsoft Point Of Service\SDK\Samples\Simulator Service Objects	A simulated Service Object binary and source code for most of the devices supported in POS for .NET. The Service Object is used with the Sample Application and the path to the Service Object is set up in the registry. The Service Object is written in C#. You will want to refer back to the source as an example for writing your own Service Object.

12. Run the sample application, TestApp.exe, found in C:\Program Files (x86)\Microsoft Point Of Service\SDK\Samples\Sample Application

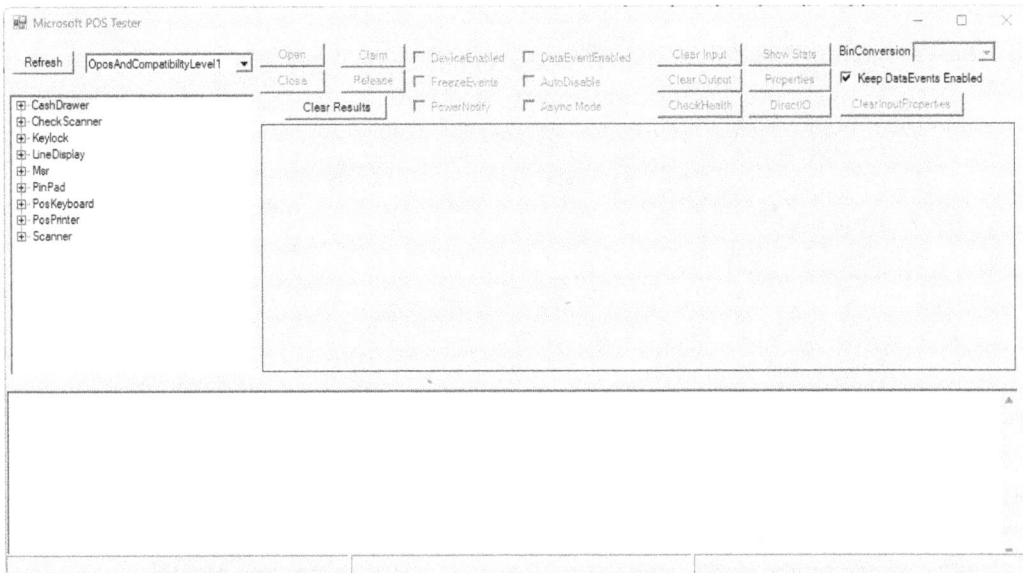

TIP: The test application is a good fallback when trying to debug POS for .NET applications and Service Objects. If you are having trouble accessing a device via your POS for .NET application, run TestApp.exe to verify that the device is set up correctly.

As you can see from the TestApp.exe, sample Service Objects and simulator devices are installed when you install the SDK. Normally, these will not be installed on a real system when only the Runtime is selected. Since we want to use the default device, we will have to address these extra devices in the development system. Now, let's see the UnifiedPOS device access sequence of open, claim, execution, release, and close in action.

13. In the Service Object tree view on the left, expand the LineDisplay device.
14. Highlight the Microsoft LineDisplay Simulator.
15. Click the Open button.
16. A dialog pops up, acting as a real LineDisplay, and the buttons for the LineDisplay controls will become available. Click the Claim button.
17. Check the DeviceEnabled button. This will enable the device for output.
18. Click the DisplayText button. The numbers in the edit box, 0123456789, will appear in the LineDisplay simulator.

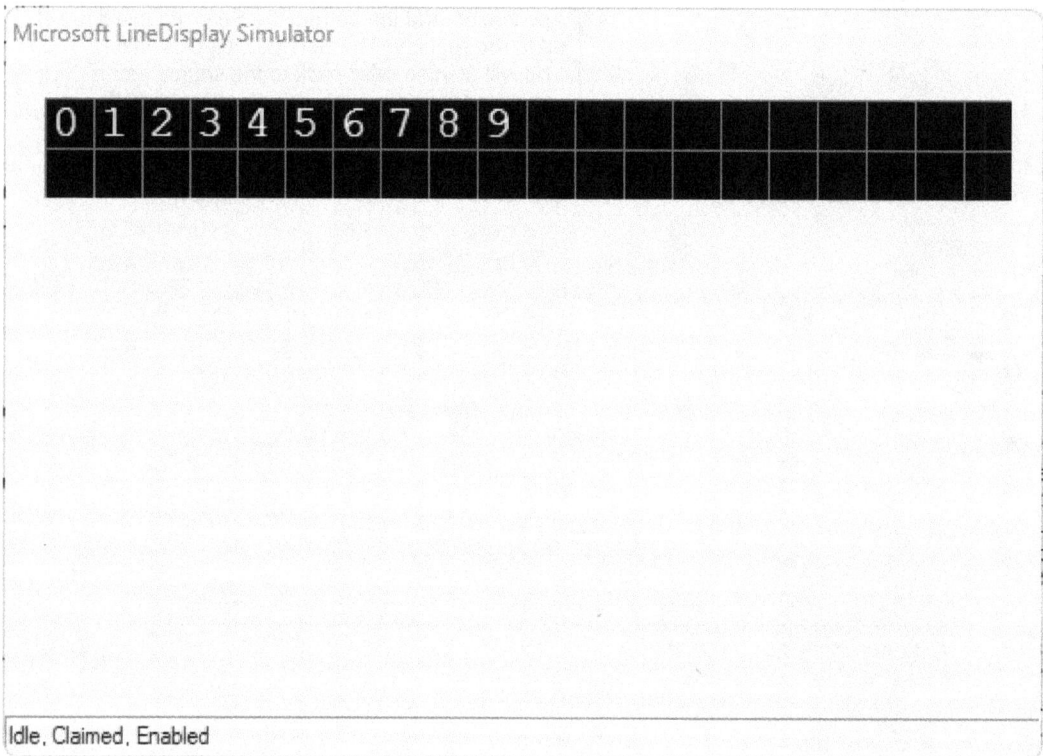

Microsoft LineDisplay Simulator

| 0 | 1 | 2 | 3 | 4 | 5 | 6 | 7 | 8 | 9 | | | | | | | | | | |

Idle, Claimed, Enabled

19. You can try the other controls to scroll text, blink the display, etc. When finished, click the Release button.
20. Then click the Close button. This will close the LineDisplay Simulator dialog.
21. Close the sample application.

Behind each of the buttons to open, claim, display text, etc., is a simple line of code that calls a specific UnfiedPOS method in the LineDisplay device class. The Testapp.exe sample application provides a good C# example of how to use the different POS device methods. We will get familiar with the POS for .NET coding in the next few chapters. The Testapp.exe application can also test some physical devices, but not all.

2.7 *Exercise 2.2 SDK: Example of Service Object Setup*

The SDK comes with an Example Service Object that supports two of the most popular POS devices: the barcode scanner and the magnetic stripe reader (MSR). In this exercise, we will see how to set up the Example Service Object for a barcode scanner and MSR. These devices will be used for developing the first POS for .NET application in the next chapter.

Hardware Requirements:

- Honeywell (HHP) Barcode Scanner – IT5600 (5600SR050) or IT3800 / Motorola (Symbol) Barcode Scanner LS2208 / or equivalent. Some barcode scanners act like keyboard devices. You want a scanner that is a pure USB HID device. Contact your POS vendor for more information.
- MagTek MSR Model #21040101 (Non-keyboard emulation model) or equivalent
- A card with a magnetic stripe (Credit card, ID card, etc.)
- Available USB ports

Software Requirements:

- POS for .NET 1.14.1
- Optional: SOManager

2.7.1 Part 1: Create an XML Configuration File for a USB HID Scanner

The Example Service Object is ready to support any USB HID barcode scanner or MSR. POS for .NET uses the PnP Hardware ID to link the hardware to the Service Object. There are two ways to create the link: either hardcoded in the Service Object or by using an XML file. For this exercise, we will do the latter. Chapter 5 will show how to hardcode the Hardware ID into the Service Object. The first thing we need to do is get the hardware ID.

1. Make sure that you have installed the POS for .NET SDK per Exercise 2.1.
2. Open Control Panel.
3. Open the System Control Panel applet.
4. Click on the Hardware tab.
5. Click on the Device Manager button.
6. With the USB scanner set to USB HID mode, plug the USB scanner device into an open USB port. Device Manager should update with two new devices listed under Human Interface Devices.

Note: Check with your Barcode scanner OEM about how to set the barcode scanner to HID mode.

7. Open the Properties for the HID-compliant Devices for the Scanner.
8. Click on the Details tab.
9. In the drop-down menu, select Hardware Ids.

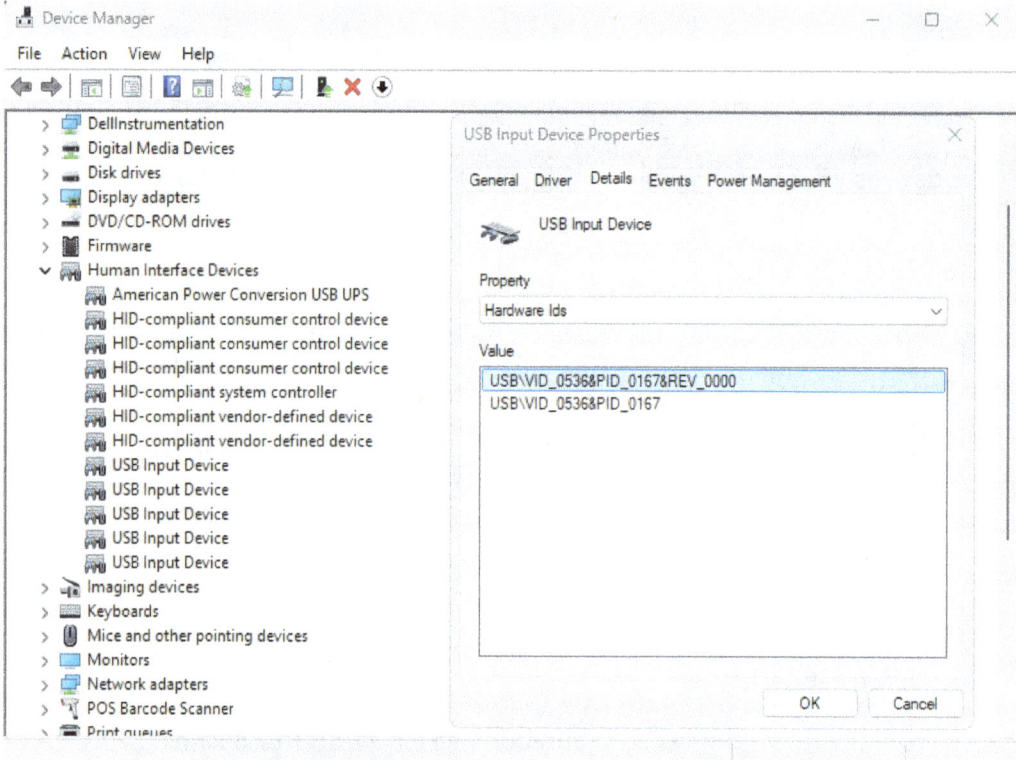

The hardware IDs for the IT5600 barcode scanner are:

    HID\Vid_0536&Pid_0167
    HID\Vid_0536&Pid_0167&Rev_0000

10. Close the HID-compliant device Properties.
11. Close Device Manager and Control Panel.
12. Open Notepad ++ or your favorite XML editor.
13. Enter the following:

```
<PointOfServiceConfig Version="1.0">
      <ServiceObject Type="Scanner" Name="Example Scanner">
            <HardwareId                  From="HID\Vid_0536&amp;Pid_0167"
To="HID\Vid_0536&amp;Pid_0167" />
            <HardwareId From="HID\Vid_0536&amp;Pid_0167&amp;Rev_0000"
To="HID\Vid_0536&amp;Pid_0167&amp;Rev_0000" />
      </ServiceObject>
</PointOfServiceConfig>
```

The ServiceObject tag defines the device type, Scanner, and the name of the Service Object, Example Scanner. If you open the ExampleScanner.cs file in Visual Studio, you will see where the name of the Service Object is set to "Example Scanner".

```
namespace Microsoft.PointOfService.ExampleServiceObjects
{
    #region ExampleScanner Class
        [ServiceObject(     DeviceType.Scanner,
                                "Example Scanner",
                    "Service Object for Example scanner", 1, 14)]
```

The <HardwareId> tag defines the HID ID range which, in this case, is the actual hardware ID for the scanner device. Since this is an XML file, the '&' must be replaced with '**&**', which is a standard XML substitution to represent the symbol '&'. The link between the device and the Service Object will fail if the change is not made.

14. Save the XML file as IT5600.xml to a folder called C:\NETPOS.
15. Copy the file from C:\NETPOS to the "C:\Program Files(x86)\Common Files\Microsoft Shared\Point Of Service\Control Configurations" directory.
16. To prove that the XML file is working, we will use a POS Device Manager utility that comes as part of the SDK. Open a command window with Administrator privileges.
17. Change the directory to "C:\Program Files(x86)\Microsoft Point Of Service".
18. With the USB scanner still plugged into a USB port, run the following command, and hit Enter:

Posdm listdevices

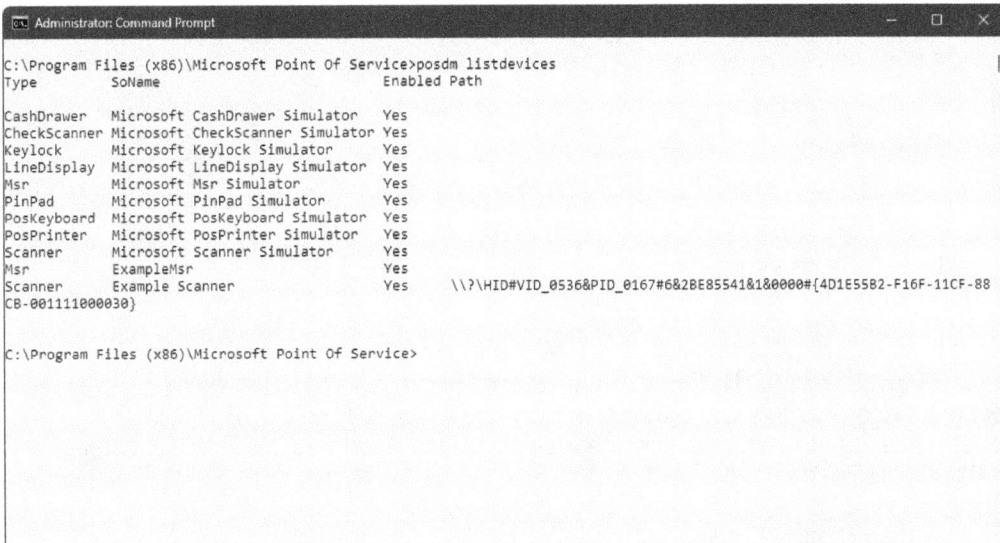

```
Administrator: Command Prompt                                          —   □   ×

C:\Program Files (x86)\Microsoft Point Of Service>posdm listdevices            |
Type        SoName                          Enabled Path

CashDrawer  Microsoft CashDrawer Simulator  Yes
CheckScanner Microsoft CheckScanner Simulator Yes
Keylock     Microsoft Keylock Simulator     Yes
LineDisplay Microsoft LineDisplay Simulator Yes
Msr         Microsoft Msr Simulator         Yes
PinPad      Microsoft PinPad Simulator      Yes
PosKeyboard Microsoft PosKeyboard Simulator Yes
PosPrinter  Microsoft PosPrinter Simulator  Yes
Scanner     Microsoft Scanner Simulator     Yes
Msr         ExampleMsr                      Yes
Scanner     Example Scanner                 Yes     \\?\HID#VID_0536&PID_0167#6&2BE85541&1&0000#{4D1E55B2-F16F-11CF-88
CB-001111000030}

C:\Program Files (x86)\Microsoft Point Of Service>
```

The output should show two scanner device types. One is the simulator Service Object and the other is the Example Scanner, which is the physical scanner connected to the system. The HID ID + GUID represents the path to the device that is used to link the application and the Service Object. An XML configuration file allows the POS device manufacturer to create one Service Object that supports multiple scanners and simply update the XML file for future scanners. Now, let's see the UnifiedPOS sequence in action with real hardware.

19. Run the TestApp.exe.
20. Select the Example Scanner from the list on the left.
21. Click the Open button.
22. Click the Claim button.
23. Check the Device Enable and DataEventEnable boxes.
24. Check the Decode Data box.
25. Use the barcode scanner to scan a barcode. The output with the device data will appear in the output window at the bottom.

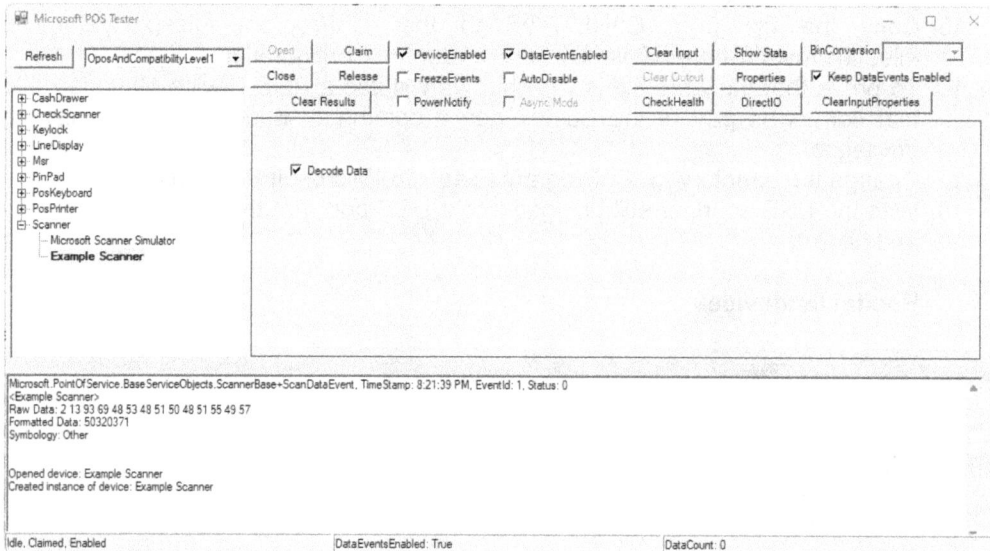

26. Click the Release and then the Close buttons when finished.
27. Close the TestApp when finished.

2.7.2   Part 2: Create an XML Configuration File for a USB HID MSR

Like the barcode scanner, the Example Service Object supports USB HID MSRs. We will perform the same steps to link the device to the Example Service Object using a configuration file.

1. Open Control Panel.
2. Open the System Control Panel applet.

24

3. Click on the Hardware tab.
4. Click on the Device Manager button.
5. Plug the MSR USB device into an open USB port. Device Manager should update with a new USB Input Device listed under Human Interface Devices.

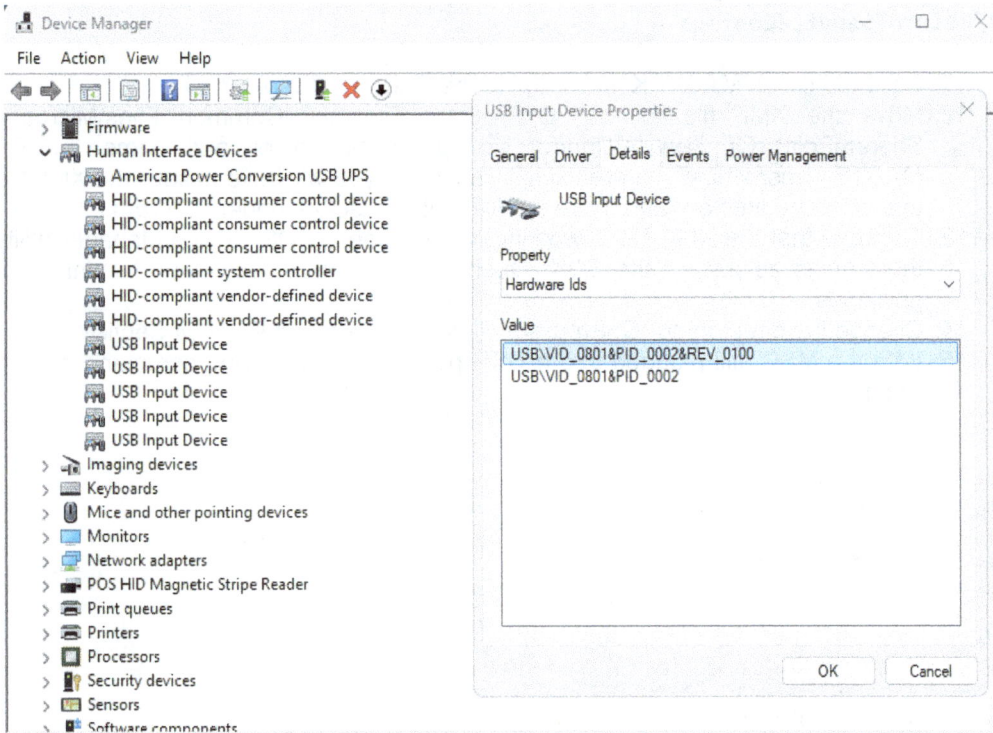

6. Open the Properties for the HID-compliant Device for the MSR.
7. Click on the Details tab.
8. In the drop-down menu, select Hardware Ids.

The hardware IDs for this example are:

```
HID\Vid_0801&Pid_0002
HID\Vid_0801&Pid_0002&Rev_0100
```

Now that we have the hardware IDs, let's create the XML file.

9. Open Notepad++.
10. Enter the following:

```
<PointOfServiceConfig Version="1.0">
    <ServiceObject Type="Msr" Name="ExampleMsr">
```

25

```
            <HardwareId                From="HID\Vid_0801&amp;Pid_0002"
To="HID\Vid_0801&amp;Pid_0002" />
            <HardwareId From="HID\Vid_0801&amp;Pid_0002&amp;Rev_0100"
To="HID\Vid_0801&amp;Pid_0002&amp;Rev_0100" />
        </ServiceObject>
</PointOfServiceConfig>
```

11. Save the file as MAGTEK_MSR.xml to the C:\NETPOS folder.
12. Move the XML file to the "C:\Program Files(x86)\Common Files\Microsoft Shared\Point Of Service\Control Configurations" folder and name the file MAGTEK_msr.xml. If you are using Windows 7, you will have to save the XML file to a different directory and move the file to the directory under Program Files.
13. To prove that the XML file is working, we will use a POS Device Manager utility that comes as part of the SDK. Open a command window with Administrator privileges.
14. Change the directory to "C:\Program Files(x86)\Microsoft Point Of Service".
15. With the MSR still plugged into a USB port, run the following command and hit Enter:

Posdm listdevices

```
Administrator: Command Prompt                                                 —  □  ×
Microsoft Windows [Version 10.0.22631.4037]
(c) Microsoft Corporation. All rights reserved.                              |

C:\Windows\System32>cd "\Program Files (x86)\Microsoft Point Of Service"

C:\Program Files (x86)\Microsoft Point Of Service>posdm listdevices
Type          SoName                                Enabled Path

CashDrawer    Microsoft CashDrawer Simulator        Yes
CheckScanner  Microsoft CheckScanner Simulator      Yes
Keylock       Microsoft Keylock Simulator           Yes
LineDisplay   Microsoft LineDisplay Simulator       Yes
Msr           Microsoft Msr Simulator               Yes
PinPad        Microsoft PinPad Simulator            Yes
PosKeyboard   Microsoft PosKeyboard Simulator       Yes
PosPrinter    Microsoft PosPrinter Simulator        Yes
Scanner       Microsoft Scanner Simulator           Yes
Scanner       Example Scanner                       Yes     \\?\HID#VID_0536&PID_0167#6&2BE85541&1&0000#{4D1E55B2-F16F-11CF-88
CB-001111000030}


C:\Program Files (x86)\Microsoft Point Of Service>
```

The output should show two MSR device types in the system. One is the simulator Service Object and the other is the Example MSR, which is the physical MSR connected to the system. The HID ID + GUID represents the path to the device that is used to link the application and the Service Object. An XML configuration file allows the POS device manufacturer to create one Service Object that supports multiple MSRs and simply update the XML file for future scanners. Now, let's see the UnifiedPOS sequence in action for real hardware.

16. Run the TestApp.exe.
17. Select the Example Msr from the list on the left.
18. Click the Open button.
19. Click the Claim button.
20. Check the Device Enable and DataEventEnable boxes.
21. Make sure the Decode Data box is checked.
22. Swipe an MSR through the MSR reader. The output with the device data will appear in the output window at the bottom. I used an old MEDC conference badge from 2006.

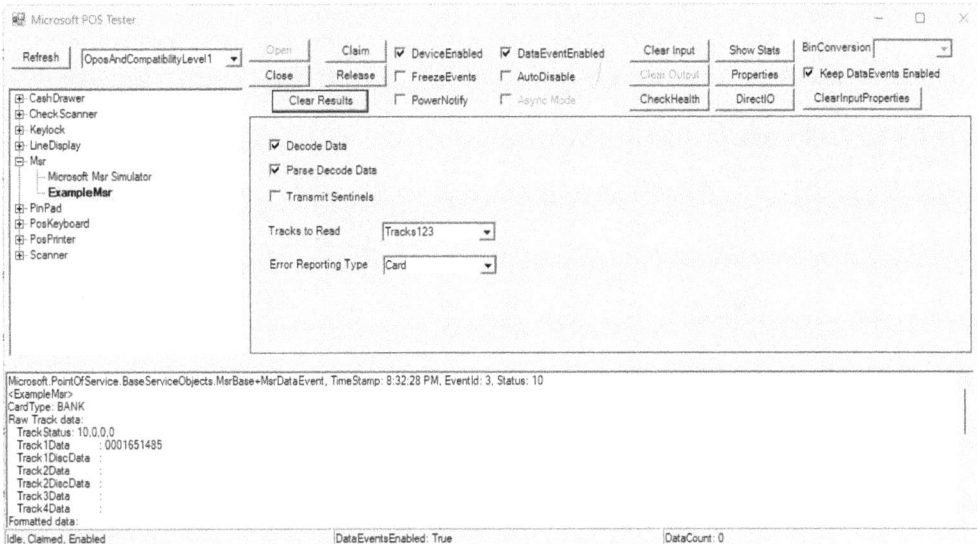

23. Click the Release and then the Close buttons when finished.
24. Close the TestApp when finished.

2.8 Summary: Ready to Go

The chapter highlighted the UnifiedPOS specification and the application sequence to access devices. The SDK's test application demonstrated the sequence with real hardware. Service Objects and OPOS drivers are the critical pieces to have in place for applications. The SDK comes with sample barcode and MSR Service Objects that work with real hardware. With the SDK installed and two POS devices set up and ready to go, we can now start building applications in the next chapter.

3 Creating POS for .NET Applications for Different POS Devices

With the barcode scanner and MSR set up in the last chapter, we can now start writing applications using the POS for .NET SDK in Visual Studio. The example applications are going to follow the "KISS" (keep it simple stupid) model for educational purposes. Rather than creating a complex application to start with, where you have to dig through hundreds of lines of code, focusing on the core API to achieve the design functionality is a good practice. Once the methods and process are understood, you can plug it into a more complex, well-written application. Since this is POS for .NET, all the applications in this chapter will be written in .NET Framework 4.8. The chapter will cover the following:

- Code example for barcode scanner
- Code example for MSR
- Code example for Pole Display
- Code example for receipt printer and cash drawer

3.1 Exercise 3.1 – Barcode Scanner Application
The barcode scanner will be used to scan a barcode and display the results in a list box.

3.1.1 Part 1: Create the Application and Form
The scanner will be used to scan barcodes and record the data in a list box. The final application will just have a list box and label.

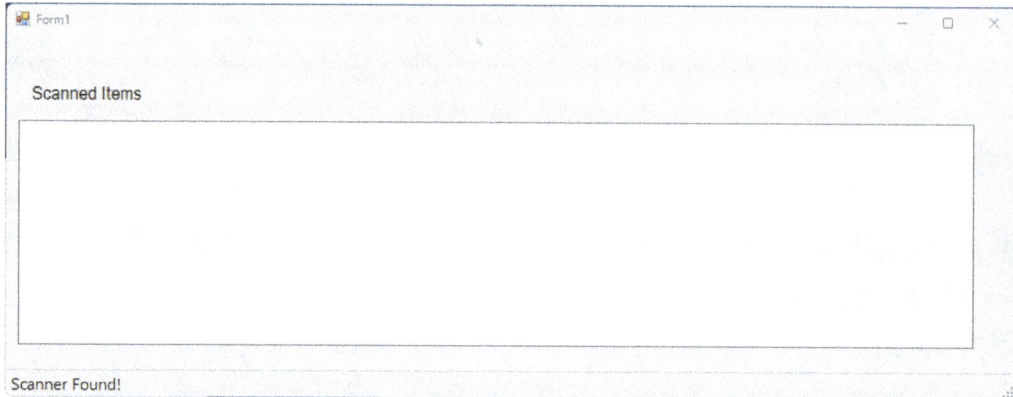

1. Open Visual Studio.
2. Click on "Create a new Project".
3. Set the language to C# and search for the "Windows Forms App (.NET Framework)" template.

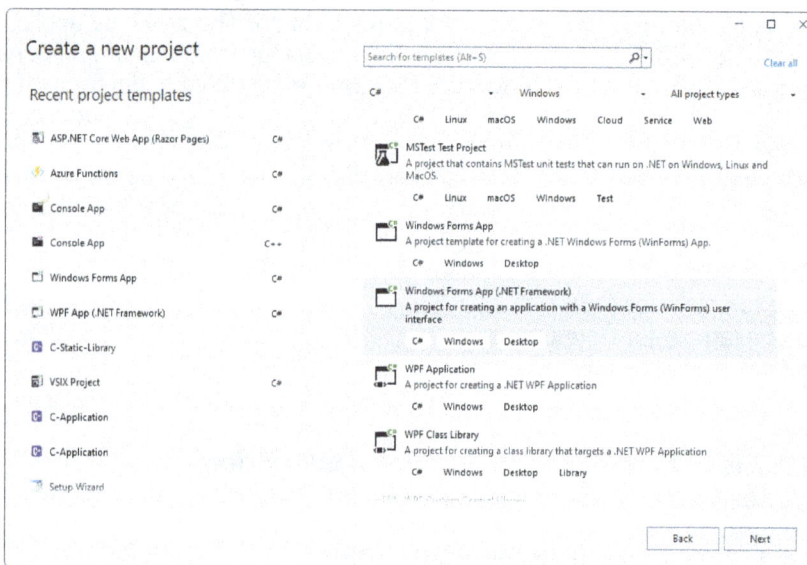

4. Click Next.
5. Name the project EX31_Barcode.
6. Set the location for the project.
7. Keep the Framework set to .NET Framework 4.8.
8. Click Create.

Configure your new project

Windows Forms App (.NET Framework)   C#   Windows   Desktop

Project name

EX31_Bar_Code

Location

C:\NETPOS\

Solution name ⓘ

EX31_Bar_Code

☐ Place solution and project in the same directory

Framework

.NET Framework 4.8

Project will be created in "C:\NETPOS\EX31_Bar_Code\EX31_Bar_Code\"

Back   Create

9. Adjust the size of Form1 to accommodate long numbers.
10. On the form, add the following controls and the properties:

    ListBox:
       Name: lstItems

    Label
       Name: lblScanItems
       Text: Scanned Items
       Font: Arial, Regular, 12pt

    StatusStrip:
       Click the drop-down to add: StatusLabel
           Name: TTStatus
           Text: Ready
           Font: Arial, Regular, 12pt

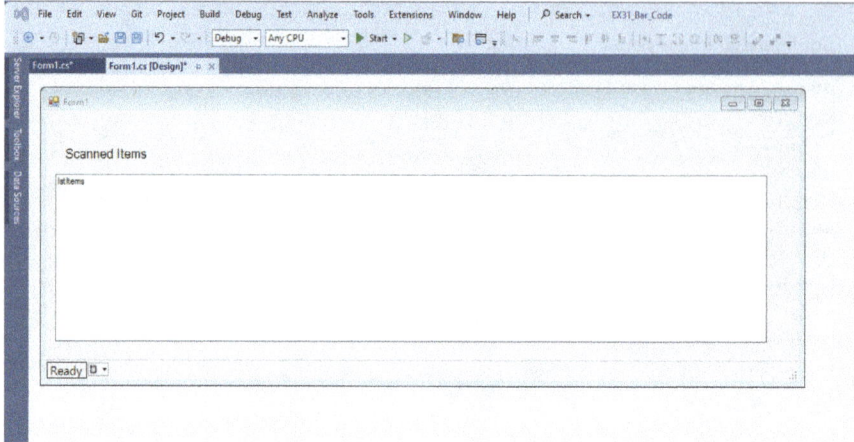

11. Save the project.

3.1.2 Part 2: Adding the POS for .NET Libraries and Code

The steps performed so far are nothing different than a typical C# application. Now, we are going to add the library reference and the code for POS for .NET. During the installation of the SDK, two libraries were installed in the "C:\Program Files(x86)\Microsoft Point Of Service\SDK" directory: Microsoft.PointOfService.dll and Microsoft.PointOfService.CotrolBase.dll. The first library will be added to our application.

1. From the menu, select Project->Add Reference. This will open the Add Reference dialog.
2. Click on the Browse tab, and locate the Microsoft.PointOfService.dll found under "C:\Program Files(x86)\Microsoft point of Service\SDK".
3. Click on the Add button.

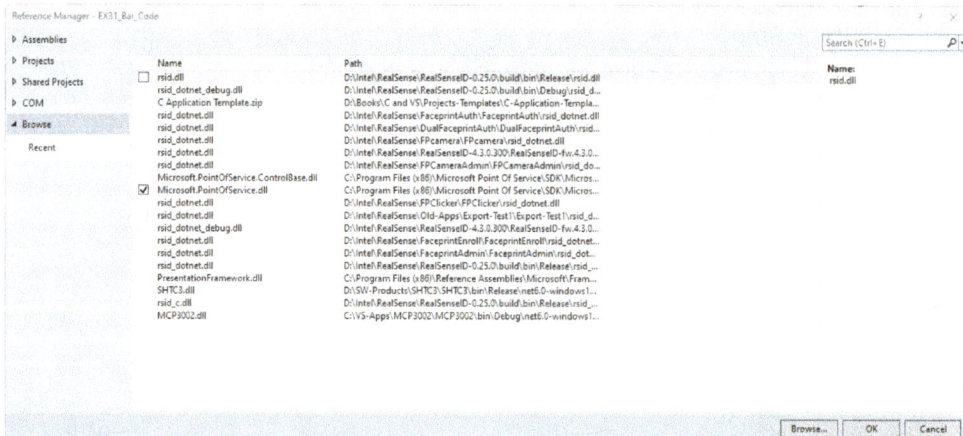

4. With the Microsoft.PointOfService.dll selected, click OK. The Microsoft.PointOfService reference is added to the project.

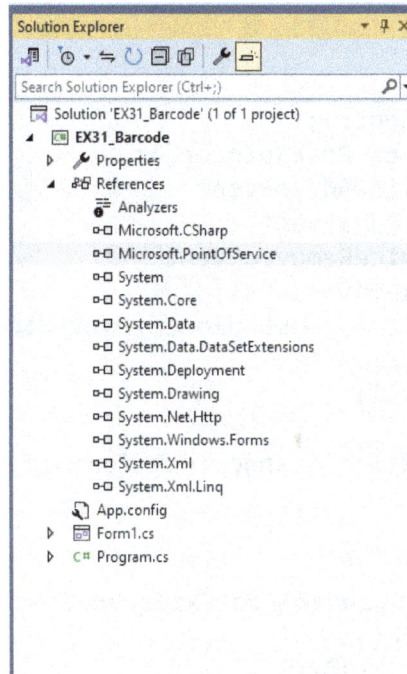

5. Open Form1.cs in code view.
6. At the top of the code before the namespace, add the Microsoft.PointOfService imports:

   using Microsoft.PointOfService

7. Within the namespace and before the class Form1, add the private global instantiation of PosExplorer and Scanner:

```
public partial class Form1 : Form
{

    private PosExplorer _posExplorer;
    private Scanner _scanner;


    public Form1()
    {
```

8. In the Form1() method, add the following after the Initalize Compoent() method call. As you type the += to set up events, just hit table and the methods for the event handlers will automatically be created after the Form1() method:

```
public Form1()
{
    InitializeComponent();
    _posExplorer = new PosExplorer();
    _posExplorer.DeviceAddedEvent                          +=
_posExplorer_DeviceAddedEvent;
    _posExplorer.DeviceRemovedEvent                        +=
_posExplorer_DeviceRemovedEvent;
    DeviceInfo device = _posExplorer.GetDevice("Scanner");

    if (device == null)
    {
        TTStatus.Text = "Scanner not found!";
    }
    else
    {
        _scanner = (Scanner)_posExplorer.CreateInstance(device);
        _scanner.Open();
        _scanner.Claim(1000);
        _scanner.DataEvent += _scanner_DataEvent;
        _scanner.DeviceEnabled = true;
        _scanner.DataEventEnabled = true;
        _scanner.DecodeData = true;
        TTStatus.Text = "Scanner Found!";
    }
}
```

The first line creates an instance of PosExplorer, which is used to get a list of all POS devices in the system. The next two lines set up the device added and removed events to address when the barcode device is plugged in or removed from a USB port. The next step is to implement the UnifiedPOS methods to open and claim the devices. The PosExplorer help file defines 5 different scenarios for creating device instances.

- Assume there is only one device of a type attached, or use only one device of a type, and default devices are configured.
- Use logical device names.
- Dynamically attach all available input devices.
- Handle all hardware configurations.
- Use a combination of these scenarios for different device types.

The code above uses the default scanner, which is going to be a problem since we have two scanners in the system: the simulator and the real scanner. We will address this a little later. If the device doesn't exist ("Nothing"), we need to post a friendly message that the scanner is not found. If the device exists, then a CreateInstance for the device is called. The Scanner will first be opened. Next, it will be claimed so no other POS application can use the device. The scanner device is enabled with a DeviceEnabled call. Finally, the scanner data events and decode are enabled.

9. Save the project.
10. Before we go on to fill in the event handlers, an invocation method needs to be added since PosExporer is not instantiated with the "this" keyword. Something glossed over from previous books is that the UI thread should be kept separate from other events like the POS for .NET runtime. The event handlers cannot directly access the UI. The solution is a background working method that can invoke the call. Add the following code after the last event handler in Form1.cs:

```
public void WriteStatus(string value)
{
    if (this.InvokeRequired)
    {
        this.Invoke(new    Action<string>(WriteStatus),    new
object[] { value });
        return;
    }
    TTStatus.Text = value;
}

public void WriteListBox(string value)
{
    if (this.InvokeRequired)
    {
        this.Invoke(new    Action<string>(WriteListBox),    new
object[] { value });
        return;
    }
    lstItems.Items.Add(value);
}
```

11. Add the following code to the posExplorer_DeviceAddedEvent method:

```
private    void    _posExplorer_DeviceAddedEvent(object    sender,
DeviceChangedEventArgs e)
{
    if (e.Device.Type == "Scanner")
```

```
     {
        _scanner                                              =
(Scanner)_posExplorer.CreateInstance(e.Device);
        _scanner.Open();
        _scanner.Claim(1000);
        _scanner.DataEvent                    +=               new
DataEventHandler(_scanner_DataEvent);
        _scanner.DeviceEnabled = true;
        _scanner.DataEventEnabled = true;
        _scanner.DecodeData = true;
        WriteListBox("Scanner attached");

     }
  }
```

The code is similar to the code used when the application starts up and looks for a barcode device. When a device is connected and the application form has focus, this routine will be called. The value "e" will hold the device type. The If-Then first determines if the device is a "Scanner". If the device is a scanner, an instance of the device will be created, and then the open and claim calls will be made and the scanner device will be enabled. Because this is a scanner, we also need to enable data events and the features to decode the data. Finally, a message will be sent to the list box via the invoke method indicating that a new scanner has been connected.

   12. Save the project.
   13. Add the following code to the _posExplorer_DeviceRemovedEvent method:

```
private   void   _posExplorer_DeviceRemovedEvent(object   sender,
DeviceChangedEventArgs e)
{
    if (e.Device.Type == "Scanner")
    {
        _scanner.DataEventEnabled = false;
        _scanner.DeviceEnabled = false;
        _scanner.Release();
        _scanner.Close();
        WriteListBox("Scanner removed");
    }

}
```

When a USB barcode scanner is unplugged, the DeviceRemovd Event disables the data event handler, disables the devices, releases the devices, and finally closes the devices. A message is sent via the invoke method that the device has been removed.

36

14. Finally, add the code for the scanner_DataEvent method:

```
private void _scanner_DataEvent(object sender, DataEventArgs e)
{
    ASCIIEncoding myEncoding = new ASCIIEncoding();
    WriteListBox(myEncoding.GetString(_scanner.ScanDataLabel));
    _scanner.DataEventEnabled = true;
}
```

The scanner returns data in bytes, so we need to use the ASCIIEncoding method to get the data in a format that we can display in a text string. The resulting text is then sent to the list box via the invoke method. By design, data events are disabled to prevent double scanning of a barcode. The last step is to re-enable data events so other scans can be made.

15. Save the project.

3.1.3   Part 3: Build and Test
An error will occur if we try to run the application with two scanners in the system. You can use POSDM to set the default device, set up a logical device name, or you can remove the simulator Service Objects. We will cover POSDM and Service Object management in the next chapter. For now, we will remove the simulator Service Object:

1. Since the registry keys are pointing to the path of the simulator Service Object, we will simply remove the Service Object. Open File Explorer and go to "C:\Program Files (x86)\Microsoft Point Of Service\SDK\Samples\Simulator Service Objects".
2. Cut and paste the Microsoft.PointOfService.DeviceSimulators.dll to the root of "C:\Program Files (x86)\Microsoft Point Of Service" folder.
3. Make sure that the XML has been created for the scanner per the last chapter, and unplug the USB scanner if it is connected to the PC.
4. Plug in the USB barcode scanner.
5. Run the "POSDM listdevices" command, and you should see that only one scanner is attached.

```
Administrator: Command Prompt                                          —  □  ×

c:\Program Files (x86)\Microsoft Point Of Service>posdm listdevices
Type      SoName          Enabled Path

Scanner   Example Scanner Yes     \\?\HID#VID_0536&PID_0167#6&2BE85541&1&0000#{4D1E5582-F16F-11CF-88CB-001111000030}

c:\Program Files (x86)\Microsoft Point Of Service>
```

6. Make sure the project is set for Debug and build the application.
7. Correct any build errors that appear.
8. Start the application by hitting F5 or selecting Debug->Start Debugging from the menu.

37

9. Once the application starts, scan a barcode from a book, soda can, water bottle, etc. The barcode should be shown on the Pole Display and the ListBox.

Note: You may not see all the numbers from the barcode. This will depend on your scanner's settings and how it handles each barcode symbology. Please see your scanner documentation for more details.

10. Unplug and plug in the scanner to verify that the add device and remove device events are working correctly.
11. Try scanning the barcodes again.

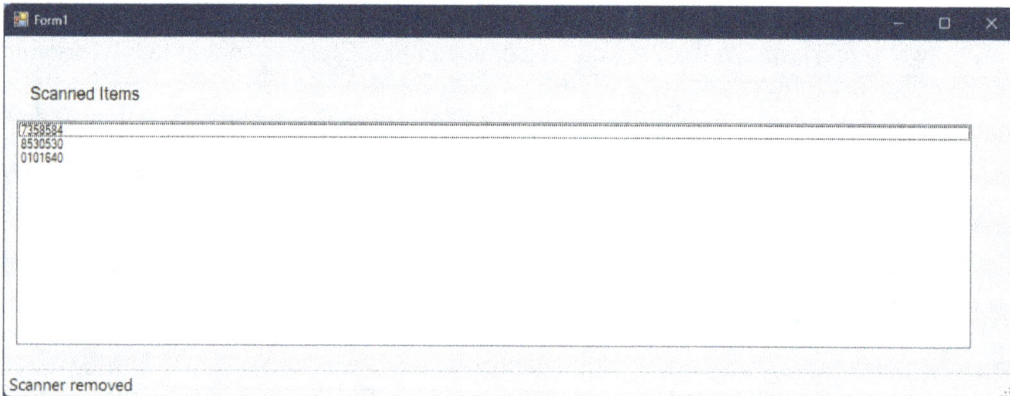

12. Stop the application when finished.

Of course, with Visual Studio, you can set breakpoints and step through the code. You can also break into and debug the Service Object source code from the application, as long as you have the source for the Service Object. We will see how this is done in the next chapter.

The exercise provides the basic example of a POS for .NET application. Plug and Play technology is an added feature since POS for .NET runs on Windows. With Plug and Play technology, your application needs to consider what happens when a device is added or

38

removed. When a device is removed, you may still want the application to run but not allow access to the features that were being used by the removed device. Like the exercises, adding DeviceAddedEvent and DeviceRemovedEvent handlers for every device is good practice.

3.2   Exercise 3.2 MSR Application

The MSR is similar to the barcode scanner; but to make things a little more interesting, the application will be a WPF .NET Framework application rather than a Forms application. Make sure that you have a USB MSR setup per the last chapter.

3.2.1   Create the Visual Studio Project

The first step is to create the project and perform the basic setup tasks.

1.   Open Visual Studio.
2.   Create a new project.
3.   Search for and select "WPF App (.NET Framework)" and click Next.

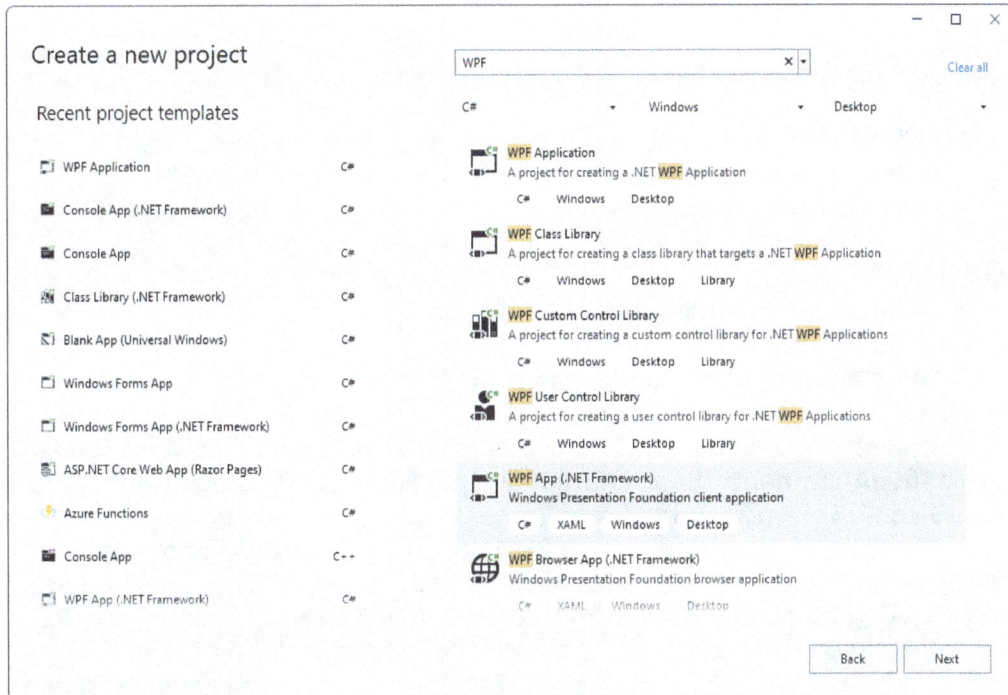

4.   Name the project EX32_MSR and click Next.
5.   Keep the .NET Framework 4.8 and click Create.

3.2.2 Set Up the XAML Controls
Now we will set up the UI.

1. In Solution Explorer, open the MainWindow.xaml.
2. Add a Label in the top, left corner.
   a. Name: lblMSRResult
   b. Font size 14 Bold
   c. Text: MSR Data
3. Add a TextBlock under the label
   a. Name: txtMSRData
   b. Text: Swipe a card
4. Add a StatusBar at the bottom of the page.
   a. Name: sBar
5. Add a TextBlock control to the status strip.
   a. Name: txtStatus
   b. Text: Ready
6. Save the project

Here is the XAML code:

```xml
<Window x:Class="EX32_MSR.MainWindow"

xmlns="http://schemas.microsoft.com/winfx/2006/xaml/presentation"
    xmlns:x="http://schemas.microsoft.com/winfx/2006/xaml"
    xmlns:d="http://schemas.microsoft.com/expression/blend/2008"
    xmlns:mc="http://schemas.openxmlformats.org/markup-compatibility/2006"
    xmlns:local="clr-namespace:EX32_MSR"
    mc:Ignorable="d"
    Title="MainWindow" Height="450" Width="800">
  <Grid>
    <Label        x:Name="lblMSRResults"      Content="MSR      Data:"
HorizontalAlignment="Left" Margin="31,68,0,0" VerticalAlignment="Top"
FontSize="14" FontWeight="Bold"/>
    <TextBlock    x:Name="txtMSRData"      HorizontalAlignment="Left"
Margin="46,144,0,0"  TextWrapping="Wrap"   Text="Swipe   a   card"
VerticalAlignment="Top" Width="700" Height="31"/>
      <StatusBar x:Name="sBar" Margin="0,378,0,0">
      <TextBlock        x:Name="txtStatus"          TextWrapping="Wrap"
Text="Ready" Width="787" Height="38"/>
    </StatusBar>
  </Grid>
</Window>
```

The Designer should look as follows:

3.2.3 Part 2: Adding the POS for .NET Libraries and Code
Next, we add the Microsoft.PointOfService.dll and the code for the MSR.

1. The first step is to add the Microsoft.PointOfService.dll reference. From the menu, select Project->Add Reference. This will open the Add Reference dialog.
2. If you previously added the reference in the last exercise, all you have to do is check the box next to Microsoft.PointOfService.dll and click OK. Otherwise, click on the Browse tab and locate the Microsoft.PointOfService.dll found under "C:\Program Files\Microsoft point of Service\SDK".
3. Click a
4. Add
5. Click on the OK button.
6. Open MainWindows.xaml.cs in code view.
7. At the top of the code, before the From1 class, add the Microsoft.PointOfService imports:

```
using System;
using System.Collections.Generic;
using System.Linq;
using System.Text;
using System.Threading.Tasks;
using System.Windows;
using System.Windows.Controls;
using System.Windows.Data;
using System.Windows.Documents;
```

41

```
using System.Windows.Input;
using System.Windows.Media;
using System.Windows.Media.Imaging;
using System.Windows.Navigation;
using System.Windows.Shapes;
using Microsoft.PointOfService;
```

8. In the Public Class MainWindow, and before the MainWindow() method, add the following global instances of PosExplorer and Msr.

```
namespace EX32_MSR
{
    /// <summary>
    /// Interaction logic for MainWindow.xaml
    /// </summary>
    public partial class MainWindow : Window
    {
        private PosExplorer myExplorer;
        private Msr myMsr;

        public MainWindow()
```

9. In the MainWindow () method, add the following code after the InitializeComponent() call. As you type the += to set up events, just hit table and the methods for the event handlers will automatically be created after the Form1() method:

```
public MainWindow()
{
    InitializeComponent();
    myExplorer = new PosExplorer();
    myExplorer.DeviceAddedEvent += MyExplorer_DeviceAddedEvent;
    myExplorer.DeviceRemovedEvent                              +=
MyExplorer_DeviceRemovedEvent;
    DeviceInfo device = myExplorer.GetDevice("Msr");

    if (device == null) {

        txtStatus.Text = "Msr not found!"''
    }
    else
    {
        myMsr = (Msr)myExplorer.CreateInstance(device);
```

```
        myMsr.Open();
        myMsr.Claim(1000);
        myMsr.DataEvent += MyMsr_DataEvent;
        myMsr.DeviceEnabled = true;
        myMsr.DataEventEnabled = true;
        myMsr.DecodeData = true;
        txtStatus.Text = "Found Msr - Ready";
    }
}
```

The first line creates an instance of PosExplorer, which is used to get a list of all POS devices in the system. The next two lines set up the device added and removed events to address when the MSR device is plugged in or removed from a USB port. The next step is to implement the UnifiedPOS methods to open and claim the devices. The PosExplorer help file defines 5 different scenarios for creating device instances.

- Assume there is only one device of a type attached, or use only one device of a type, and default devices are configured.
- Use logical device names.
- Dynamically attach all available input devices.
- Handle all hardware configurations.
- Use a combination of these scenarios for different device types.

The code above uses the default MSR. Hopefully, you have moved the example simulator Service Object DLL out of the way per the previous exercise. If the device doesn't exist ("Nothing"), we need to post a friendly message that the line display is not found. If the device exists, then a CreateInstance for the device is called. The MSR will first be opened. Next, it will be claimed so no other POS application can use the device. The MSR device is enabled with a DeviceEnabled call. Finally, the MSR data events and decode are enabled.

10. Add the following code to the MyExplorer_DeviceAddedEvent method:

```
private    void    MyExplorer_DeviceAddedEvent(object    sender,
DeviceChangedEventArgs e)
{
    if (e.Device.Type == "Msr")
    {
        myMsr = (Msr)myExplorer.CreateInstance(e.Device);
        myMsr.Open();
        myMsr.Claim(1000);
        myMsr.DataEvent                         +=                         new
DataEventHandler(MyMsr_DataEvent);
        myMsr.DeviceEnabled = true;
```

43

```
        myMsr.DataEventEnabled = true;
        myMsr.DecodeData = true;
        Dispatcher.Invoke((Action)(() => txtStatus.Text = "MSR
connected"));
    }
}
```

The code performs the same MSR open and claim steps as the MainWindow() method. Since this is a thread, the Displatcher.Invoke call is require to communicate with the TextBlock in the UI thead.

11. Save the project.
12. Add the following code to the MyExplorer_DeviceRemovedEvent method:

```
private   void   MyExplorer_DeviceRemovedEvent(object   sender,
DeviceChangedEventArgs e)
{
    if (e.Device.Type == "Msr")
    {
        myMsr.DecodeData = false;
        myMsr.DataEventEnabled = false;
        myMsr.DeviceEnabled = false;
        myMsr.Release();
        myMsr.Close();
        Dispatcher.Invoke((Action)(() => txtStatus.Text = "MSR
removed"));
    }
}
```

13. Save the project
14. Add the following code to the MyMSR_DataEvent method:

```
private void MyMsr_DataEvent(object sender, DataEventArgs e)
{
    ASCIIEncoding myEncoding = new ASCIIEncoding();
    Dispatcher.Invoke((Action)(()    =>    txtMSRData.Text    =
myEncoding.GetString(myMsr.Track2Data)));
    myMsr.DataEventEnabled = true;
}
```

Like the Scanner, the MSR Base Class disables further events. The last step is to re-enable data events so other scans can be made.

15. Save the project.

44

3.2.4   Part 3: Build and Test

1. Make sure you have the MSR XML configuration file set up per Chapter 2 exercise.
2. Make sure the Simulator Service Object is moved to a different location per the last exercise. You can check this by plugging in the MSR and running "posdm.exe listdevices" to check that the ExampleMsr is the only Service Object device available.

```
Administrator: Command Prompt                                             —     □
Microsoft Windows [Version 10.0.22631.4037]
(c) Microsoft Corporation. All rights reserved.

C:\Windows\System32>cd "\Program Files (x86)\Microsoft Point Of Service"

C:\Program Files (x86)\Microsoft Point Of Service>posdm listdevices
Type      SoName     Enabled Path

Msr       ExampleMsr Yes     \\?\HID#VID_0801&PID_0002#6&12D29640&0&0000#{2A9FE532-0CDC-44F9-9827-76192F2CA2FB}


C:\Program Files (x86)\Microsoft Point Of Service>
```

3. Make sure the project is set for Debug, and build the application.
4. Start the application by hitting F5 or selecting Debug->Start Debugging from the menu.
5. The status label should display that an MSR is attached. Swipe a card with a magnetic strip. The data should be shown in the text box.

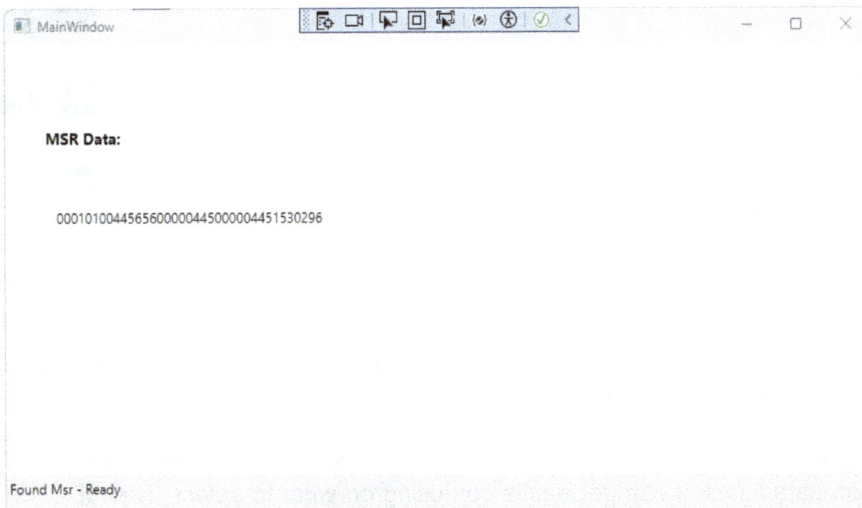

```
MainWindow                                                        —     □     ×


    MSR Data:


        0001010044565600000445000004451530296






Found Msr - Ready
```

If you don't see any data, data might be on another track. Change the myMsr.Track2Data to the different tracks. There are up to 4 tracks available.

6. Test the device remove and add events.
7. When finished stop debugging

3.3 Epson OPOS ADK for .NET v1.14.30E Setup

Epson Pole Display DM-D110 and Epson Receipt Printer TM-T88V will be used for the next three exercises. As mentioned in the previous chapter, choosing the right hardware to work with POS for .NET is important. Specifically, an OPOS driver or Service Object is needed. Epson supplies an OPOS driver specific for POS for .NET, and configuring the OPOS driver for use with these devices is the critical step. This section covers installing the Epson OPOS ADK for these two POS devices.

Note: There are many pole-displays and receipt printers on the market. The only reason these two devices are called out in the books is that I actually have the devices to test with.

1. Go to the Epson.com site.
2. Search for one of the Epson devices. In my case, I searched on DM-D110.
3. At the bottom of the product page for the DM-D110, there is a filter to select an OS. I selected Windows 11. Any version of Windows will do.
4. Click Go and a list of support files is presented.
5. Expand the Utilities section to find the Epson OPOS ADK for .NET v1.14.30E.

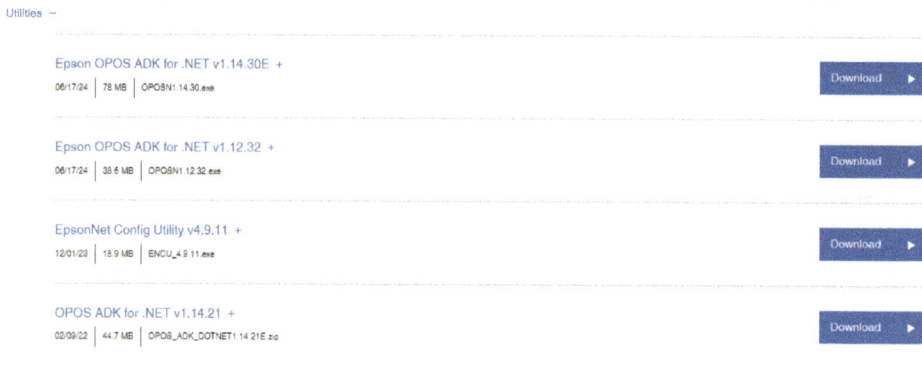

Utilities −

Epson OPOS ADK for .NET v1.14.30E +
06/17/24 | 78 MB | OPOSN1.14.30.exe   Download ▶

Epson OPOS ADK for .NET v1.12.32 +
06/17/24 | 38.6 MB | OPOSN1.12.32.exe   Download ▶

EpsonNet Config Utility v4.9.11 +
12/01/23 | 18.9 MB | ENCU_4.9.11.exe   Download ▶

OPOS ADK for .NET v1.14.21 +
02/09/22 | 44.7 MB | OPOS_ADK_DOTNET1.14.21E.zip   Download ▶

6. Click the Download button next to Epson OPOS ADK for .NET v1.14.30E. This will download the OPOSN1.14.30.exe.

Note: There are other drivers available, but these are for different programming and device interaction paradigms. It can get a little confusing on what to select. The OPOS ADK for .NET v1.14.30E is the only item needed for POS for .NET.

7. Run the installer

8. The installer wizard runs and there are two screens that are important to the installation. The first is the type of setup. Select Custom.
9. Change the installation location to some location other than the default Program File folder.
10. Click Next.

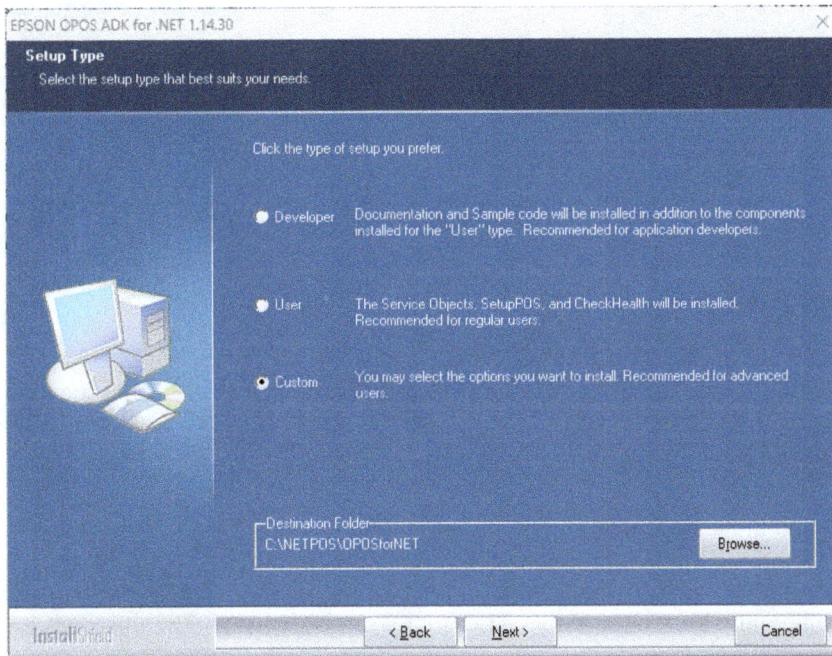

11. The next screen is the advanced settings screen. For the communication ports to be used, select Serial, USB, and Ethernet. The pole-display will use serial and the receipt printer supports both USB and Ethernet. Make sure Serial, USB, and Ethernet are select keep all the other items selected. Click Next.

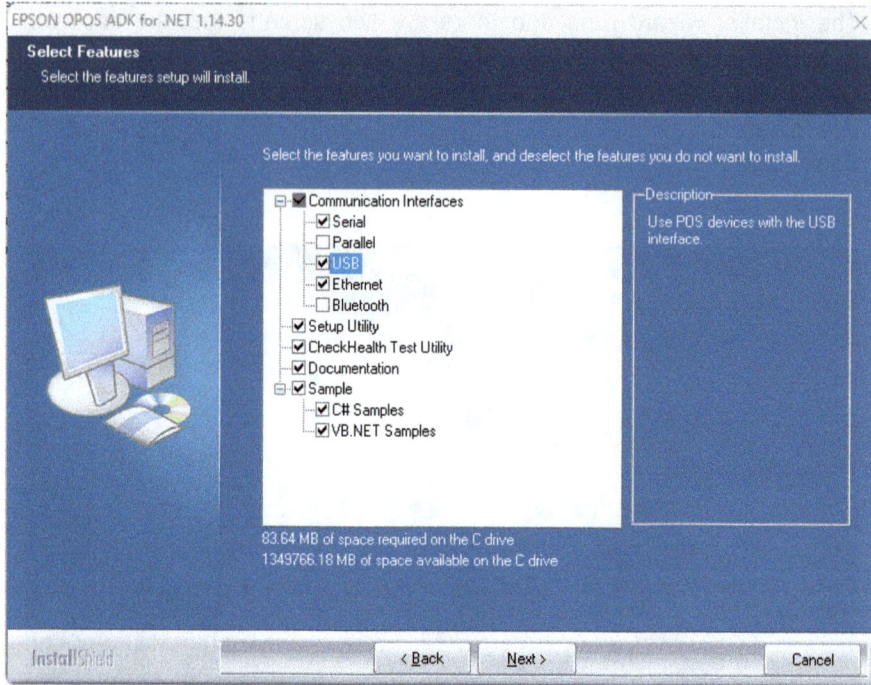

12. Finish the installation.

The start menu will contain shortcuts under the Epson OPOS ADK for .NET folder for documentation and the Epson Setup POS for OPOS.NET utility. The utility will be used to configure the OPOS driver.

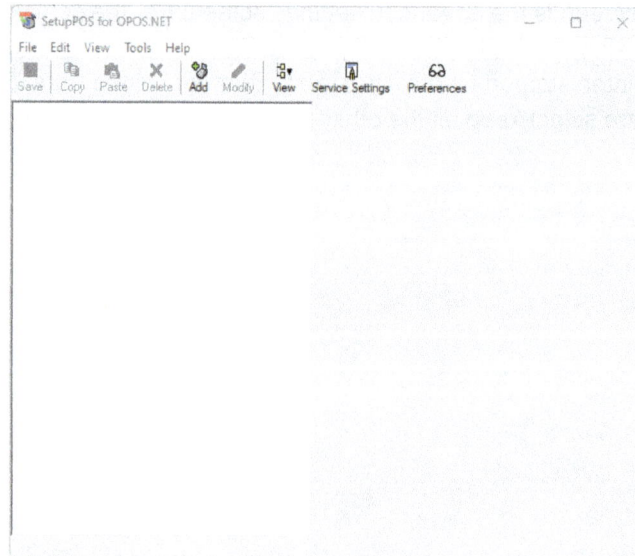

Warning: The Epson OPOS ADK for .NET v1.14.30E installs the OPOS driver to support almost all Epson POS devices. Running POSDM listdevices will result in a long list of available devices. Logical Device Names will be needed to filter out the specific device that the application will connect to. Too many unused devices are not ideal, and can be a bit confusing. This is how the software support for POS for .NET was made. The newer Epson OPOS ADK that we will use for POS WinRT has a better solution, but it doesn't work with POS for .NET.

3.4   Exercise 3.3 Pole Display Application

The Pole Display application for this exercise will have a text box for user input and a button that will send the text to the Pole Display. The OPOS driver needs to be configured. The DM-D110 can connect via serial port as a standalone device or via USB port connected to the Receipt Printer. My TM-T88V has the UB-E04 Connection-It™ module that supports Ethernet. The DM-D110 requires the UB-U01III or U02III to connect to the printer. For this book, the standalone serial connection will be used.

Note: Pole Display and Line Display are used interchangeably. The imply the same device.

3.4.1   Part 1: Set Up the OPOS driver for the Pole Display
The first step is to set up the OPOS driver for the line display.

1. Connect the Pole Display to the computer via serial port or USB-to-serial. Also, make sure the power switch under the DM-D110 is set to on.
2. Run the Epson Setup POS for OPOS.NET utility with administrator rights.
3. Click the Add button from the toolbar.
4. The first thing to do is to click on the Device Category and select LineDisplay.
5. Select the Device Name as DM-D110.
6. The Logical Device Name can be set to anything, but leave it as LineDisplay.

Note: The utility interface is not the best. A more intuitive design would have the drop-down first followed by the Device Name. The Logical Device Name comes last, as it can be changed by the user.

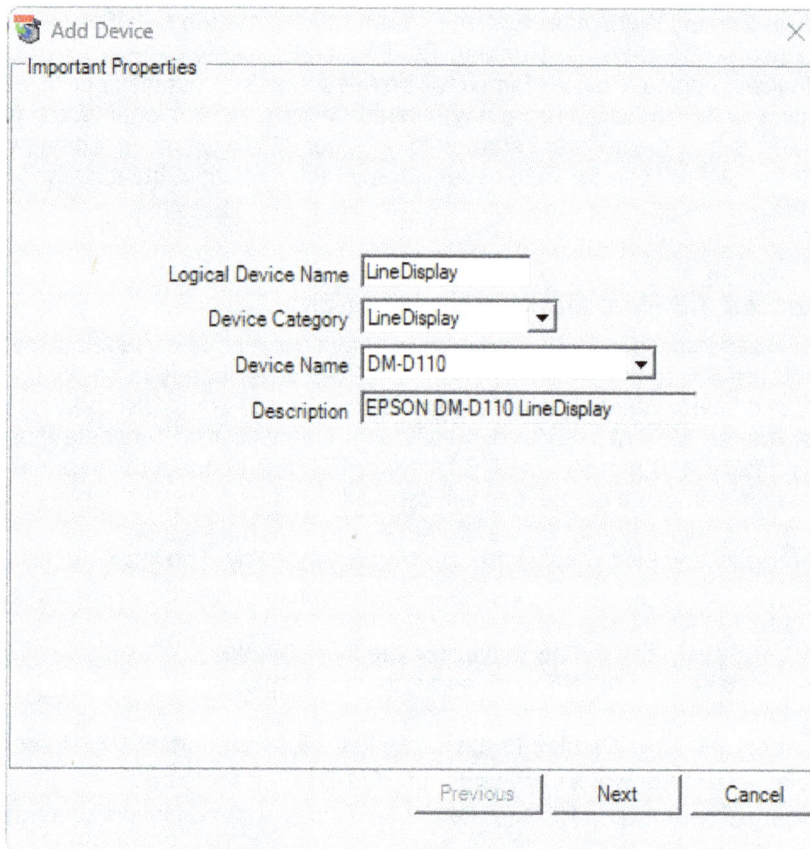

7. Click Next.
8. In the next screen set the port type to Serial.
9. A USB-to-serial connector can be used to connect to a PC without serial ports. Looking in Device Manager, the COM port is COM4. Change the Port Name to COM4.

```
Add Device                                              ×
┌─ Connection Interface ──────────────────────────────────┐
│                                                         │
│              Port Type  Serial          ▼               │
│                                                         │
└─────────────────────────────────────────────────────────┘

┌─ Serial Specific Settings ──────────────────────────────┐
│                                                         │
│              Port Name  COM4                            │
│        Output Buffer Size  4096                         │
│         Bits per second  9600 bps    ▼                  │
│              Data bits  8 Bits   ▼                      │
│                 Parity  No Parity Bit    ▼              │
│              Stop Bits  1 Stop Bit   ▼                  │
│           Flow Control  DTR/DSR   ▼                     │
│                                                         │
│                                                         │
│                   Previous     Next      Cancel         │
└─────────────────────────────────────────────────────────┘
```

10. Click Next.
11. Set the connection Type to Stand Alone.

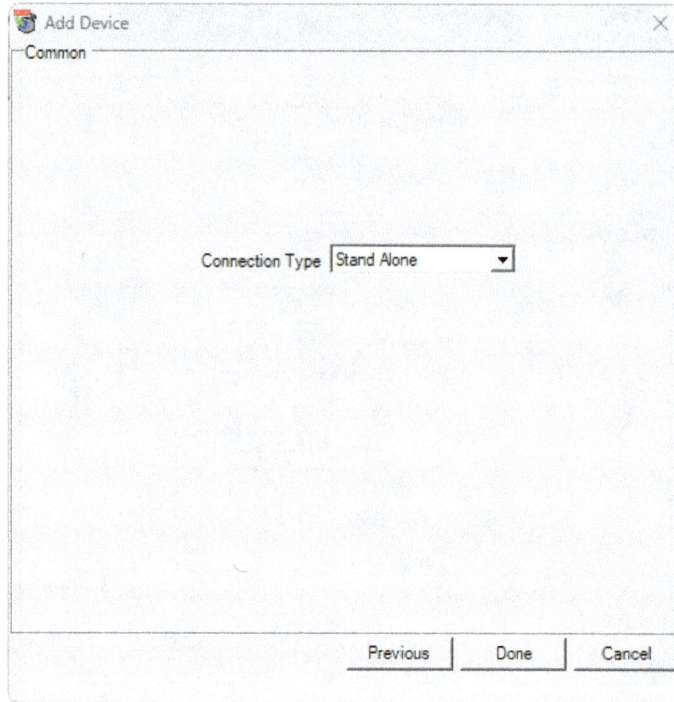

12. Click Next.
13. The LineDisplay appears in the list. Click Save so the changes can be saved for the OPOS driver configuration.

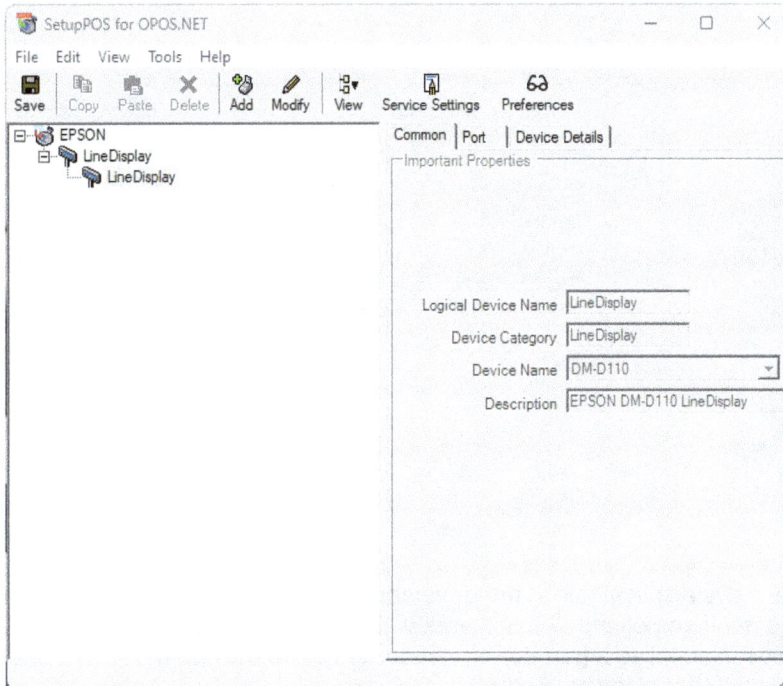

14. The next step is to make sure the OPOS driver is working correctly. This is where the TestApp.exe application comes in handy. Run the TestApp.exe application that is part of the POS for .NET SDK.
15. Expand the device branch on the left for LineDisplay.
16. You will see a device called LineDisplay. This I the logical name for the DM-D110 that was defined when we configured the OPOS driver. Click on LineDisplay.
17. Click Open button
18. Click Claim
19. Check the box for DeviceEnabled.
20. There is a "Display Text" button and next to the button is a text box containing a message to send to the display. Click the "Display Text" button. The numbers should appear on the display. If it doesn't, go back and check the Epson Setup POS for OPOS.NET utility to make sure the OPOS driver is configured correctly.

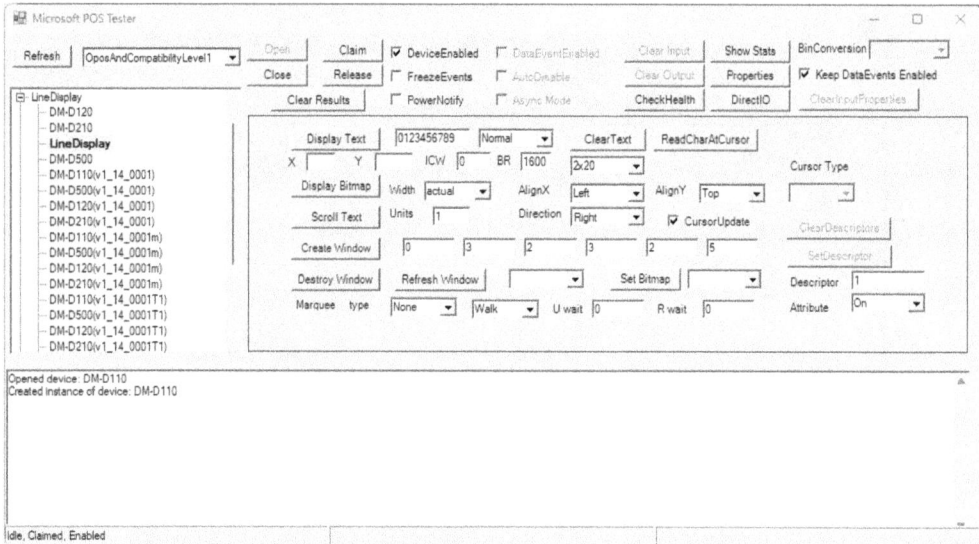

TestApp.exe plays a critical role in the development and debug setup of OPOS drivers and Service Objects. If something doesn't work with your application, use the TestApp.exe as a fallback. Epson supplies a number of Visual Studio projects for their POS device as part of the ADK, which are also very helpful.

3.4.2   Part 2: Create the Pole Display Application
With the OPOS driver configured and tested, the application development can proceed.

1. Open Visual Studio.
2. Click on "Create a new Project".
3. Set the language to C# and search for "Windows Forms App (.NET Framework)" template:

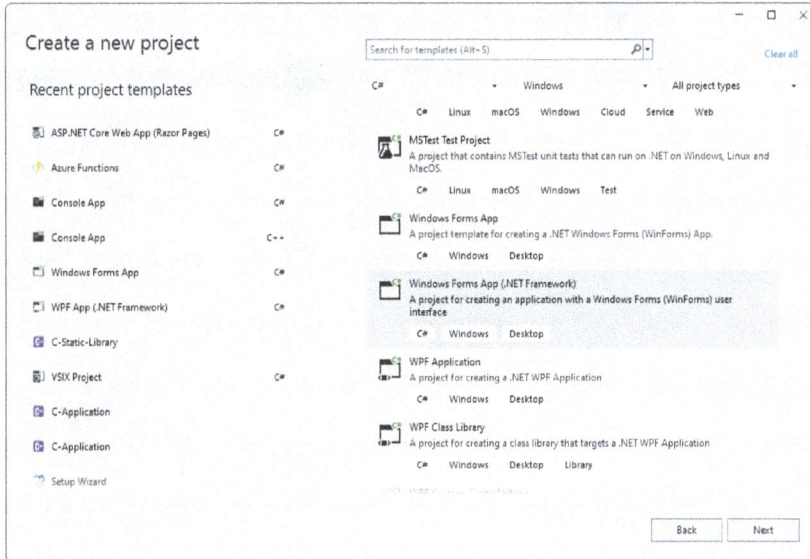

4.   Click Next.
5.   Name the project EX33_PoleDisplay.
6.   Set the location for the project.
7.   Keep the Framework set to .NET Framework 4.8.
8.   Click Create.
9.   Adjust the size of Form1 to accommodate long numbers.
10. On the form, add the following controls and the properties:

        TextBox:
            Name: txtInput
            Font: Arial, Regular, 12pt

        Label:
            Name: lblPD
            Text: Input some text:
            Font: Arial, Regular, 12pt

        Button:
            Name: btnSend
            Text: Send to Display
            Font: Arial, Regular, 12pt

        StatusStrip:
            Click the drop-down to add: StatusLabel
                Name: TTStatus
                Text: Ready
                Font: Arial, Regular, 12pt

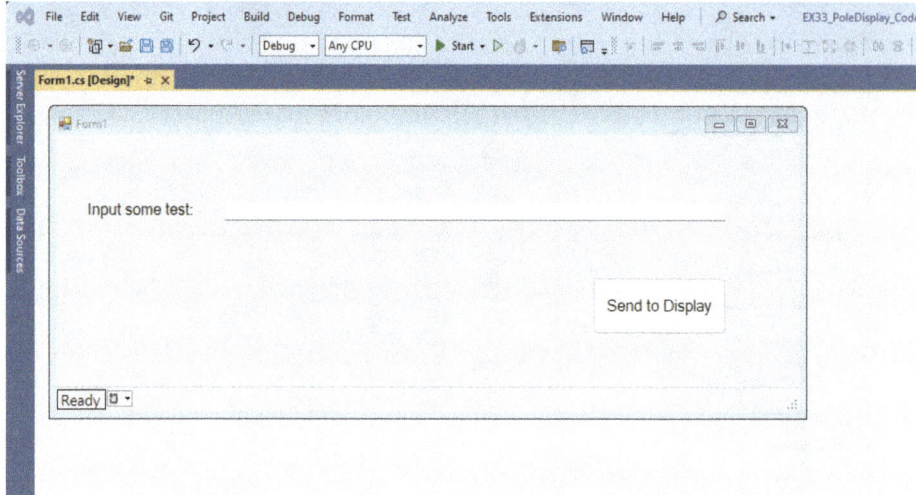

11. Save the project.

3.4.3 Part 3: Add the POS for .NET Libraries and Code
Next, we add the Microsoft.PointOfService.dll and the code for the Pole Display.

1. The first step is to add the Microsoft.PointOfService.dll reference. From the menu, select Project->Add Reference. This will open the Add Reference dialog.
2. If you previously added the reference in the last exercise, all you have to do is check the box next to Microsoft.PointOfService.dll and click OK. Otherwise, click on the Browse tab, and locate the Microsoft.PointOfService.dll found under "C:\Program Files\Microsoft point of Service\SDK".
3. Click add.
4. Click on the OK button.
5. Open Form1.CS in code view.
6. At the top of the code, before the From1 class, add the Microsoft.PointOfService imports:

```
using System;
using System.Collections.Generic;
using System.ComponentModel;
using System.Data;
using System.Drawing;
using System.Linq;
using System.Text;
using System.Threading.Tasks;
using System.Windows.Forms;
using Microsoft.PointOfService;
```

56

7. Before the Public Class Form1, add the following global instances of PosExplorer and LineDisplay:

```
namespace EX33_PoleDisplay
{
    public partial class Form1 : Form
    {

        public PosExplorer myPosExplorer;
        public LineDisplay myLineDisplay;

        public Form1()
```

8. In the Form1() method, add the following code after the InitializeComponent() call:

```
public Form1()
{
    InitializeComponent();
    myPosExplorer = new PosExplorer();
    DeviceInfo                        device                        =
myPosExplorer.GetDevice(DeviceType.LineDisplay,"LineDisplay");

    if (device == null) {

        TTStatus.Text = "Pole Display not found";
    }
    else
    {
        myLineDisplay                                              =
(LineDisplay)myPosExplorer.CreateInstance(device);
        myLineDisplay.Open();
        myLineDisplay.Claim(1000);
        myLineDisplay.DeviceEnabled = true;
         TTStatus.Text = "Pole Display found - Ready";
    }
}
```

Since a logical device name has been assigned to the PoleDisplay, the GetDevice method calls out the logical device name to get the device. Unlike the barcode scanner and MSR, the Pole Display connects via serial port, which is not a Plug and Play device. DeviceAdded and DeviceRemoved events are not needed. Also, DateEvent is not required since data is not being capture. The Pole Display is a very simple device to setup.

9. Go back to the Form1.cs Designer.
10. Double-click on the button control so the button handler code is added to Form1.cs.
11. Add the following code for the button handler:

```
private void btnSend_Click(object sender, EventArgs e)
{
    try
    {
        myLineDisplay.DisplayText(txtInput.Text,
DisplayTextMode.Normal);
    }
    catch(PosControlException)
    {
        TTStatus.Text = "Display error";
    }
}
```

When the button is pressed, any text in the text box will be set to the line display.

3.4.4 Part 4: Build and Test
With the code ready to go, let's test the application.

1. Make sure the Pole Display is connected and the OPOS driver has been configured.
2. Make sure the project is set for Debug and build the application.
3. Start the application by hitting F5 or selecting Debug->Start Debugging from the menu.
4. The status label should display that a new scanner is attached. Enter some text and click the "Send to Display" button. The text should appear on the display.

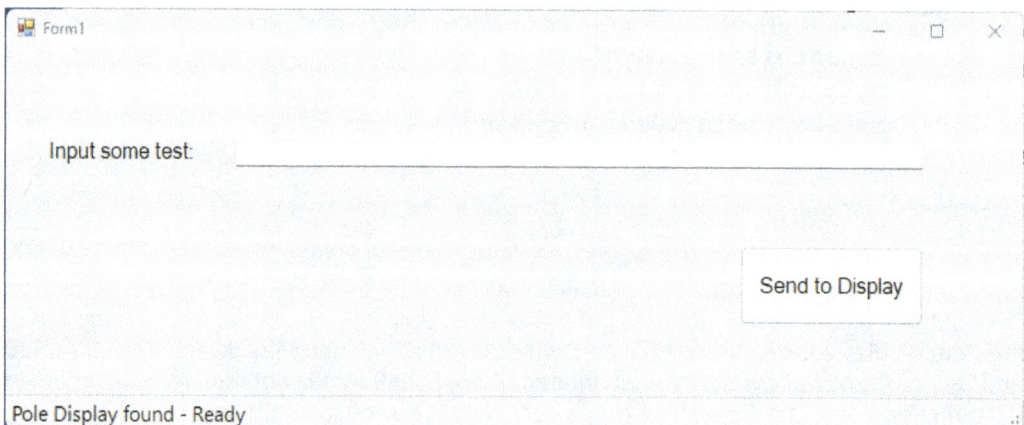

The Pole Display is a simple device that the application opens and claims and never has to release and close. A Release and Close could be performed on application exit, but since the myLineDisplay object will be destroyed, the additional code will be redundant.

3.5   Exercise 3.4 Receipt Printer Application

The USB connection will be used for this application. We will save connecting via Ethernet for a later exercise in the book. This application will look similar to the Pole Display where a push of a button sends text from a text box to a printer instead of the Pole Display.

3.5.1   Part 1: Set Up the OPOS driver for the Receipt Printer

For my printer, there are two communication options requiring two steps to set up the printer and OPOS driver.

1. The first step starts with the printer hardware. With the system powered off and paper in the printer, power on the printer while pressing and holding the Feed button.
2. The printout shows a menu selection using the Feed button as the input. Press and hold the Feed button and a menu list of settings, 0 to 14, print out on the paper. Press and hold the Feed button for one second and then press and hold to enter the system settings. For my printer, the USB interface is setting 11. Hit the Feed button 10 times and then press and hold on the 11th. The printer shows the USB interface options and which port is the default (marked with '*'). Perform the press-and-hold trick again to change the interface to the built-in USB. The print out says the built-in USB is selected and then goes back to the main menu.
3. Turn off the power when finished.
4. Connect the Receipt Printer to the computer.
5. Power on the printer. You should hear the USB connection sound from the computer.
6. Run the Epson Set Up POS for OPOS.NET utility with administrator rights.
7. Click the Add button from the toolbar.
8. Change the Logical Device Name to epsonPrinter.
9. Change the Device Name to TM-T88V.
10. Click Next.

11. Change the Port Type to USB.
12. Click Next.

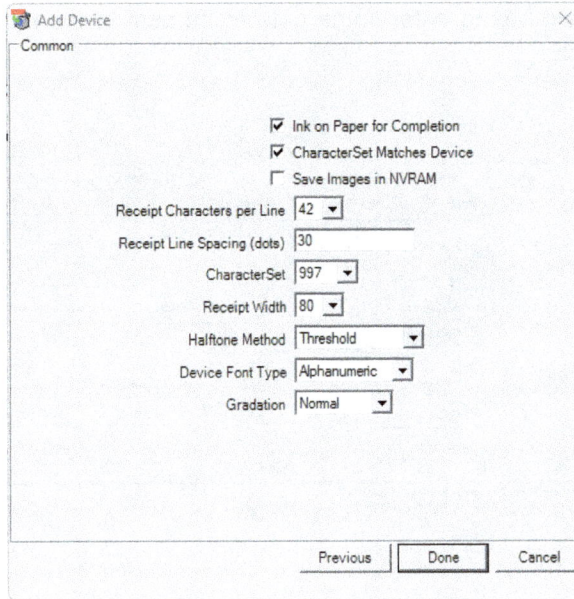

13. Keep the defaults on the next screen and click Done. The utility shows the Receipt Printer and the Pole Display.

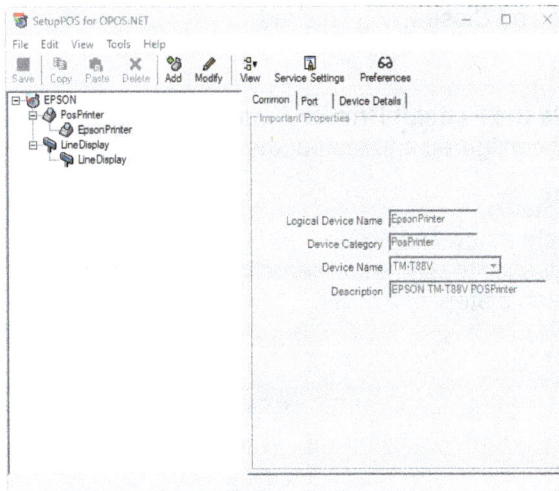

14. Run the TestApp.exe from the POS for .NET SDK.
15. Select the epsonPrinter from the device list on the left under the PosPRinter branch.
16. Click Open.
17. Click Claim.
18. Check the box for DeviceEnabled.

61

19. There is a text box to enter some data to be sent to the printer. Click the Print Normal button.

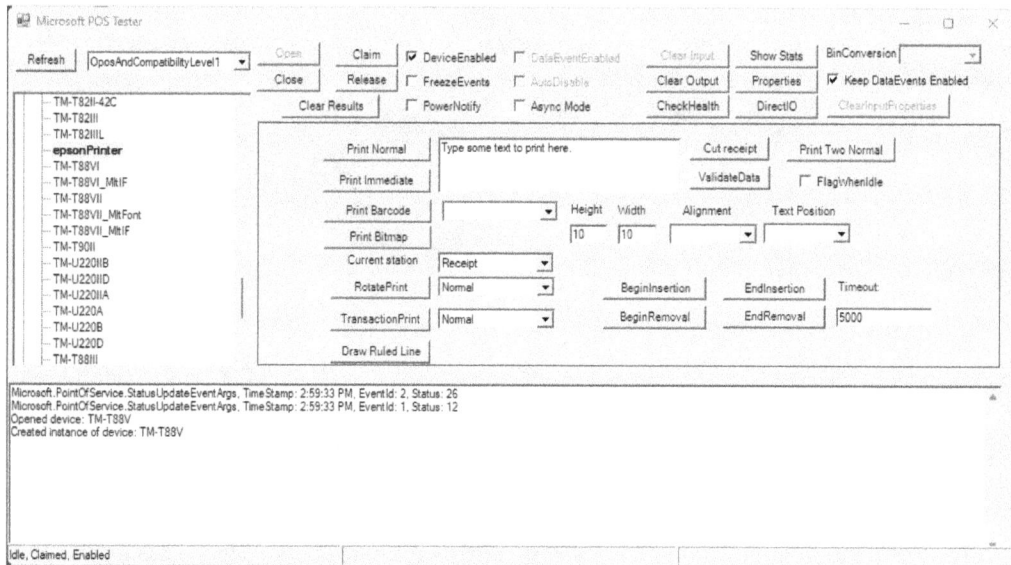

20. On the printer, click the Feed button a few times to scroll up the message.
21. Click Release and Close when finished.

3.5.2 Part 2: Create the Receipt Printer Application

With the OPOS driver configured and tested, the application development can proceed.

1. Open Visual Studio.
2. Click on "Create a new Project".
3. Set the language to C# and search for the "Windows Forms App (.NET Framework)" template:

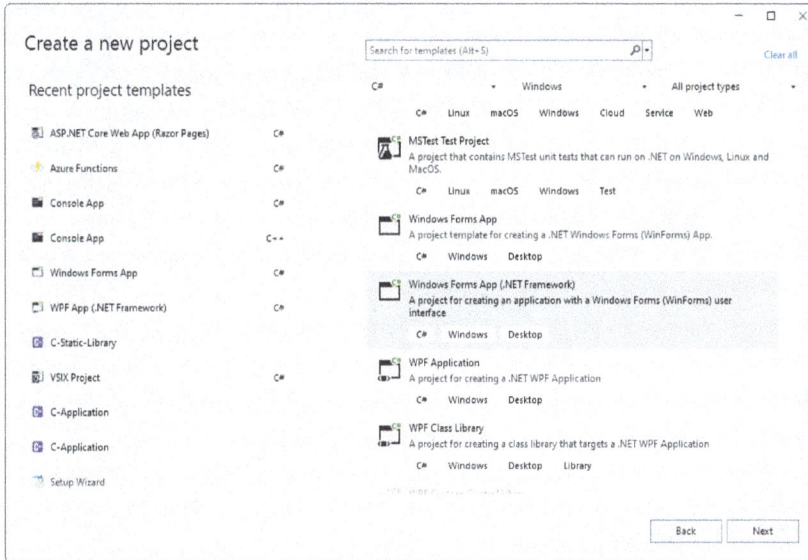

4. Click Next.
5. Name the project EX34_Printer.
6. Set the location for the project.
7. Keep the Framework set to .NET Framework 4.8.
8. Click Create.
9. Adjust the size of Form1 to accommodate long numbers.
10. On the form, add the following controls and their properties:

> TextBox:
> > Name: txtInput
> > Font: Arial, Regular, 12pt

> Label:
> > Name: lblPrint
> > Text: Input some text:
> > Font: Arial, Regular, 12pt

> Button:
> > Name: btnSend
> > Text: Send to Printer
> > Font: Arial, Regular, 12pt

> StatusStrip:
> > Click the drop-down to add: StatusLabel
> > > Name: TTStatus
> > > Text: Ready
> > > Font: Arial, Regular, 12pt

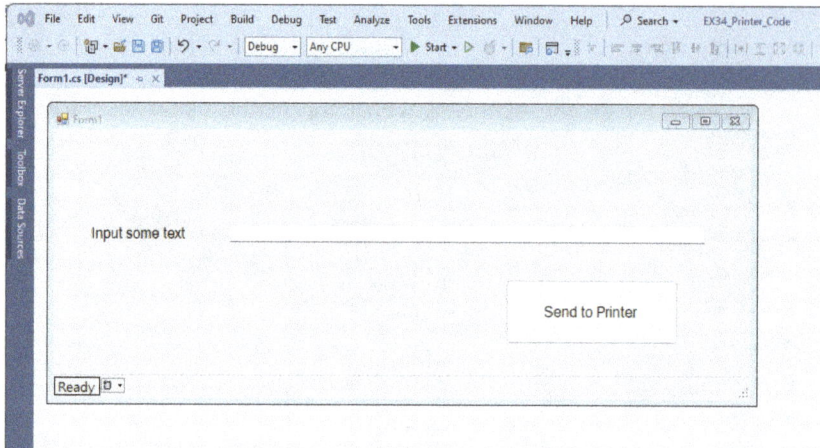

11. Save the project.

3.5.3 Part 3: Add the POS for .NET Libraries and Code
Next, we add the Microsoft.PointOfService.dll and the code for the Pole Display.

1. The first step is to add the Microsoft.PointOfService.dll reference. From the menu, select Project->Add Reference. This will open the Add Reference dialog.
2. If you previously added the reference in the last exercise, all you have to do is check the box next to Microsoft.PointOfService.dll and click OK. Otherwise, click on the Browse tab, and locate the Microsoft.PointOfService.dll found under "C:\Program Files\Microsoft point of Service\SDK".
3. Click add.
4. Click on the OK button.
5. Open Form1.CS in code view.
6. At the top of the code before the From1 class, add the Microsoft.PointOfService imports:

```
using System;
using System.Collections.Generic;
using System.ComponentModel;
using System.Data;
using System.Drawing;
using System.Linq;
using System.Text;
using System.Threading.Tasks;
using System.Windows.Forms;
using Microsoft.PointOfService;
```

7. Before the Public Class Form1, add the following global instances of PosExplorer and PosPrinter:

```
namespace EX34_Printer
{
    public partial class Form1 : Form
    {
        public PosExplorer myPosExplorer;
        public PosPrinter myPOSPrinter;

        public Form1()
```

8. In the Form1() method, add the following code after the InitializeComponent() call. As you type the += to set up events just hit table and the methods for the event handlers will automatically be created after the Form1() method:

```
public Form1()
{
    InitializeComponent();
    myPosExplorer = new PosExplorer();
    myPosExplorer.DeviceAddedEvent                              +=
MyPosExplorer_DeviceAddedEvent;
    myPosExplorer.DeviceRemovedEvent                           +=
MyPosExplorer_DeviceRemovedEvent;
    DeviceInfo                          device                    =
myPosExplorer.GetDevice(DeviceType.PosPrinter, "epsonPrinter");

    if (device == null) {

        TTStatus.Text = "Printer not found!";
    }
    else
    {
        myPOSPrinter                                             =
(PosPrinter)myPosExplorer.CreateInstance(device);
        myPOSPrinter.Open();
        myPOSPrinter.Claim(1000);
        myPOSPrinter.DeviceEnabled = true;
        TTStatus.Text = "Printer found - Ready";
    }
}
```

The code follows the previous exercises with a PosExplorer instance created and DeviceAdded and Device Removed events. The Receipt Printer device instance is created using the logical device name that was entered during the OPOS driver setup. The typical open, claim, and device enable are called.

9. Create a method to handle the event handler's output to the UI thread:

```
public void WriteStatus(string value)
{
    if (this.InvokeRequired)
    {
        this.Invoke(new      Action<string>(WriteStatus),      new
object[] { value });
        return;
    }
    TTStatus.Text = value;
}
```

10. Add the code to the MyPosExplorer_DeviceAddedEvent() handler method:

```
private    void    MyPosExplorer_DeviceAddedEvent(object    sender,
DeviceChangedEventArgs e)
{
    myPOSPrinter
=(PosPrinter)myPosExplorer.CreateInstance(e.Device);
    myPOSPrinter.Open();
    myPOSPrinter.Claim(1000);
    myPOSPrinter.DeviceEnabled=true;
    WriteStatus("Printer added");
}
```

11. Add the code to the MyPosExplorer_DeviceRemovedEvent() handler method:

```
private   void   MyPosExplorer_DeviceRemovedEvent(object    sender,
DeviceChangedEventArgs e)
{
    myPOSPrinter.Release();
    myPOSPrinter.Close();
    WriteStatus("Printer removedd");
}
```

12. Switch to the Form1.cs Designer and double-click on the button to add the button handler code to the Form1.cs.

13. Add the code to the button handler:

```
private void btnSend_Click(object sender, EventArgs e)
{
    //Send the text the printer the \n is required to complete
the print and line feed.

myPOSPrinter.PrintNormal(PrinterStation.Receipt,txtInput.Text+"
\n");

    //scroll the paper forward a bit
    myPOSPrinter.PrintNormal(PrinterStation.Receipt, "\n");
    myPOSPrinter.PrintNormal(PrinterStation.Receipt, "\n");
    myPOSPrinter.PrintNormal(PrinterStation.Receipt, "\n");
    myPOSPrinter.PrintNormal(PrinterStation.Receipt, "\n");
    myPOSPrinter.PrintNormal(PrinterStation.Receipt, "\n");

    //Cut the paper 90%
    myPOSPrinter.CutPaper(90);
}
```

When the button is clicked, the information that has been entered into the text box will be sent to the printer. The "\n" is needed for the print to be executed. If the "\n" was missing, the printer would hold the string in the buffer until an escape or carriage return would be sent. Several blank lines are sent to scroll up the paper before the paper is cut to 90%. When designing receipts, you will have to address the layout of the data.

3.5.4   Part 4: Build and Test

Everything is normal so far with this application compared to the other applications. There are two problems with this application, however, and running the test will demonstrate what they are.

1. Make sure the Receipt Printer is connected and the OPOS driver has been configured.
2. Make sure the project is set for Debug and build the application.
3. Start the application by hitting F5 or selecting Debug->Start Debugging from the menu.
4. Bang! An invoke error based on an unstable stack is hit with the DeviceEnable call.

```
26
27    v         if (device == null) {
28
29                 TTStatus.Text = "Printer not found!";
30             }
31    v         else
32             {
33                 myPOSPrinter = (PosPrinter)myPosExplorer.CreateInstance(device);
34                 myPOSPrinter.Open();
35                 myPOSPrinter.Claim(1000);
36                 myPOSPrinter.DeviceEnabled = true;  ⊗
37                 TTStatus.Text = "Printer found – Read";
38             }
39         }
40
          1 reference
41    v    private void MyPosExplorer_DeviceRemovedEve
42        {
43            myPOSPrinter.Release();
44            myPOSPrinter.Close();
45            WriteStatus("Printer removedd");
46        }
47
          1 reference
48    v    private void MyPosExplorer_DeviceAddedEvent
49        {
50            myPOSPrinter =(PosPrinter)myPosExplorer
```

Exception Thrown   ▶ ⏸ ✕

Managed Debugging Assistant 'PinvokeStackImbalance' : 'A call to
PInvoke function 'Epson.opos.tm.service.v1_14_0001!
jp.co.epson.upos.core.v1_14_0001.pntr.init.Verifiability::GetRandomData
1' has unbalanced the stack. This is likely because the managed
PInvoke signature does not match the unmanaged target signature.

🔷 Ask Copilot | Show Call Stack | Copy Details | Start Live Share session

▲ Exception Settings
  ☑ Break when this exception type is thrown
     Except when thrown from:
     ☐ EX34_Printer_Code.exe
  Open Exception Settings | Edit Conditions

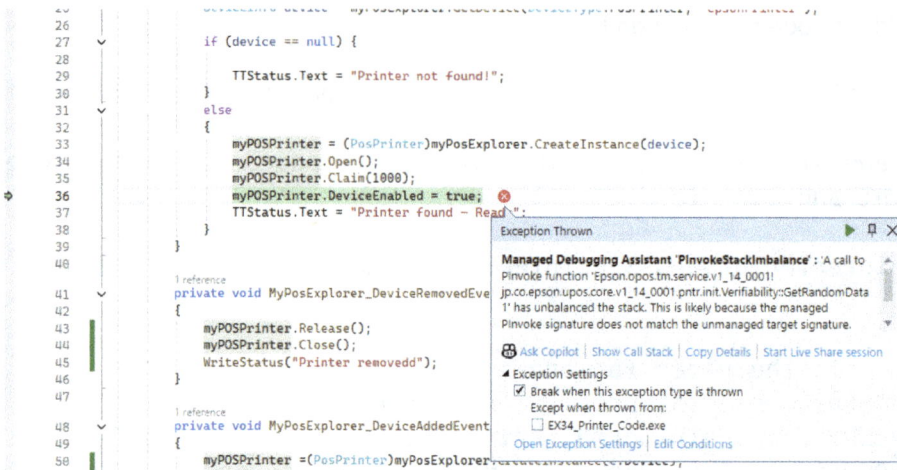

Since the source code to the OPOS driver is not available, one could bounce around Internet searches and find useless information. The issue is that the application is being built as a 32-bit app, which is running on a 64-bit OS. A change to the project build is in order.

5. Stop the debugger.
6. From the menu, go to Project->EX34_printer_code properties.
7. Go to the Build section.
8. Uncheck the Prefer 32-bit selection.

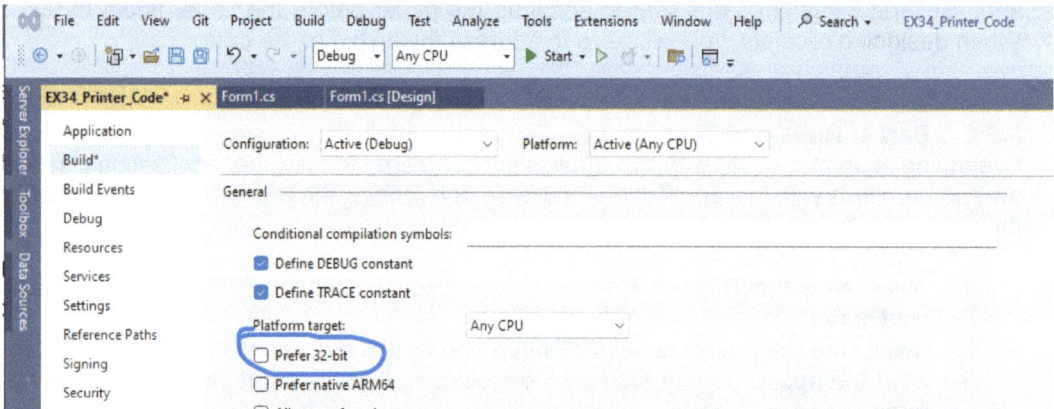

9. Save the project.
10. Run the debugger again. This time the error is avoided.
11. Enter some text in the text box and hit the button to send the data to the printer. The printer should print the message, scroll up a bit, and cut the paper.

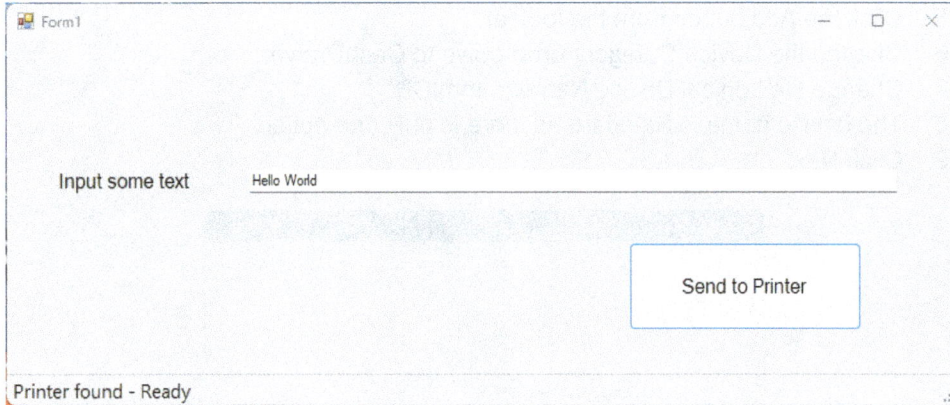

12. Now, let's test the DeviceAdded and DeviceRemoved events. Set a breakpoint in the MyPosExplorer_DeviceRemovedEvent handler method.
13. Unplug the USB Receipt Printer. Nothing happens.
14. Plug the USB Receipt Printer back in. Nothing happens other than the Windows USB connection sound.
15. Change the text in the text box and hit the button to send the data to the printer. The printer still prints.
16. Stop debugging when finished.

The DeviceAdded and DeviceRemoved events were never called. The second issue is that the OPOS driver doesn't handle plug and play events like the Service Object. The DeviceAdded and DeviceRemoved event code is useless in the application for this OPOS driver. The Plug and Play is irrelevant for an Ethernet connection as well.

3.6 Exercise 3.5 Adding the CashDrawer Device to the Printer Application

The MMF CashDrawer that I am using is very basic. The Cash Drawer device connects to the DK (device kick) port of the Receipt Printer. A single-method call is made to simply open the Cash Drawer device. There are more sophisticated cash drawers with USB interfaces and OPOS drivers or Service Objects that can provide data like cash drawer open/close status.

3.6.1 Part 1: Setup the OPOS driver for the Cash Drawer
Now let's modify the Epson OPOS driver to add cash drawer support.

1. Connect the Receipt Printer to the computer.
2. Connect the cash drawer to the DK port of the Receipt Printer.
3. Power on the printer. You should hear the USB connection sound in Windows.
4. Run the Epson Setup POS for OPOS.NET utility with administrator rights.

69

5. Click the Add button from the toolbar.
6. Change the Device Category drop-down to CashDrawer.
7. Change the Logical Device Name to mmfCD.
8. The device name is Standard as there is only one option.
9. Click Next

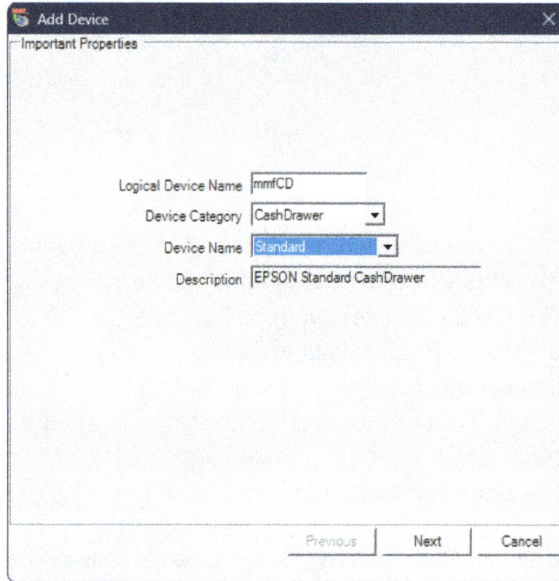

10. Set the connection interface to USB.
11. Set the connection port to the epsonPrinter.
12. Click Next.

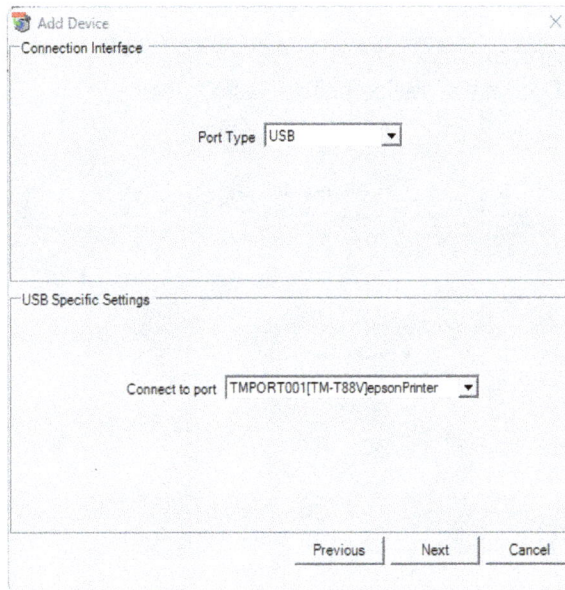

13. Keep the defaults on the next screen and click Done. The CashDrawer is now added to the list.

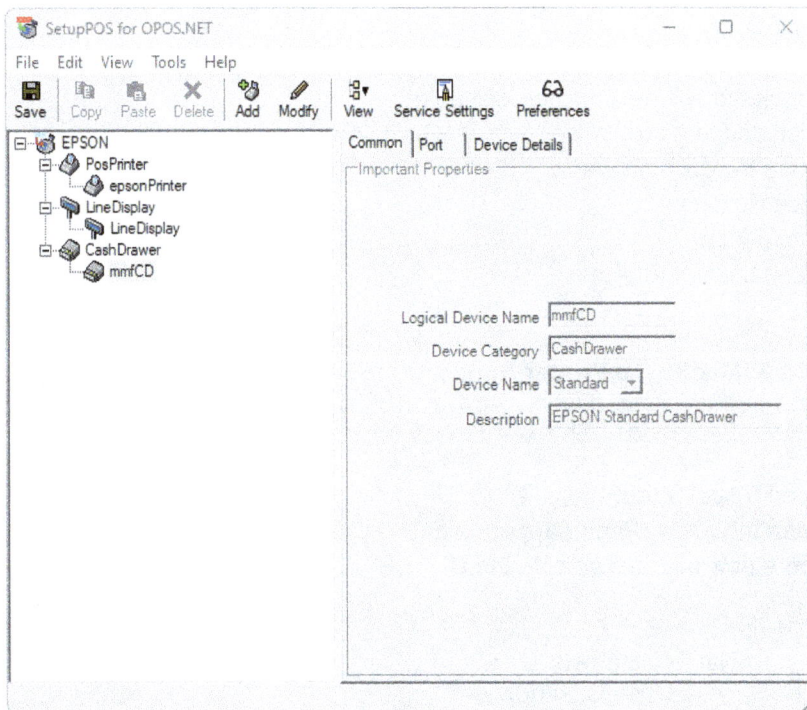

14. Click Save.
15. Run the TestApp.exe.
16. Select mmfCD from the device under CashDrawer.

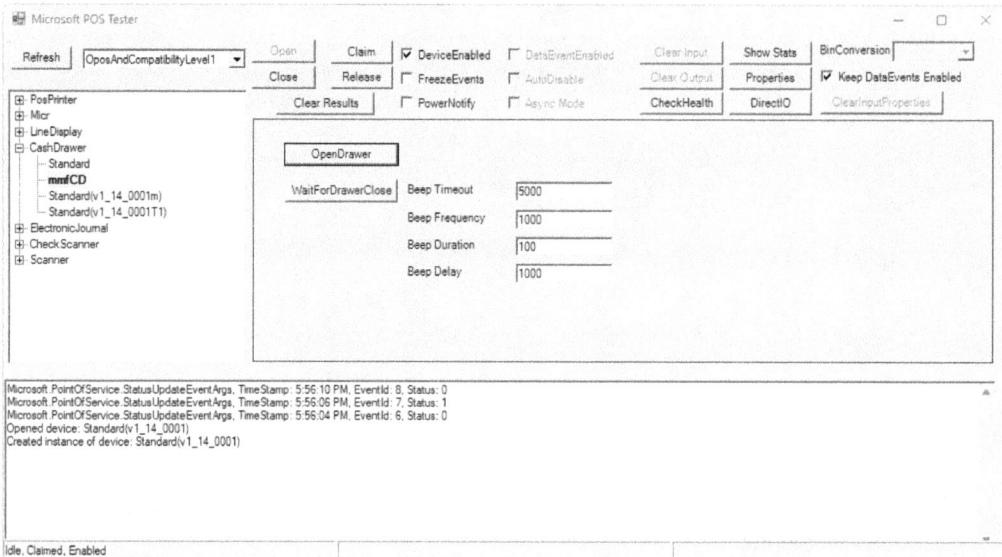

17. Click Open.
18. Click Claim.
19. Check the box next to DeviceEnabled.
20. Click on the OpenDrawer button, which should open the cash drawer.
21. Close the cash drawer.
22. Click Release.
23. Click Close.
24. Close the test app.

3.6.2   Part 2: Modify the Printer Application

There is no need to create a whole new application since the cash drawer is tied to the Receipt Printer. Let's modify what we already have.

1. Open Visual Studio.
2. Open the EX34_Printer application.
3. Add a new button control to the UI:

        Button:
            Name: btnOpenCD
            Text: Open Cashdrawer
            Font: Arial, Regular, 12pt

4. Double-click on the button to create the button handler method.
5. Before we add code to the handler, we need to create a CashDrawer instance and create the device. After the PosExplorer and PosPrinter declarations before Form1() method and the declaration for the CashDrawer:

```
namespace EX34_Printer
{
    public partial class Form1 : Form
    {
        public PosExplorer myPosExplorer;
        public PosPrinter myPOSPrinter;
        public CashDrawer myCashdrawer;

        public Form1()
```

6. In the Form1() method add the code to

```
public Form1()
{
    InitializeComponent();
    myPosExplorer = new PosExplorer();
    myPosExplorer.DeviceAddedEvent                              +=
MyPosExplorer_DeviceAddedEvent;
    myPosExplorer.DeviceRemovedEvent                           +=
MyPosExplorer_DeviceRemovedEvent;
    DeviceInfo                        device                       =
myPosExplorer.GetDevice(DeviceType.PosPrinter, "epsonPrinter");

    if (device == null) {

        TTStatus.Text = "Printer not found!";
    }
    else
    {
        myPOSPrinter                                              =
(PosPrinter)myPosExplorer.CreateInstance(device);
        myPOSPrinter.Open();
        myPOSPrinter.Claim(1000);
        myPOSPrinter.DeviceEnabled = true;
        TTStatus.Text = "Printer found";
```

```
    }

    DeviceInfo                          deviceCD                    =
myPosExplorer.GetDevice(DeviceType.CashDrawer, "mmfCD");

    if (deviceCD == null) {

        TTStatus.Text += "Cash Drawer not found!";
    }
    else
    {
        myCashdrawer                                                =
(CashDrawer)myPosExplorer.CreateInstance(deviceCD);
        TTStatus.Text += " Cash Drawer found - Ready";
    }
}
```

This time the open, claim, and device enable calls are not made when the cash drawer instance is created. This will be saved for the button handler.

7. Add the code to the new button handler method:

```
private void btnOpenCD_Click(object sender, EventArgs e)
{
    myCashdrawer.Open();
    myCashdrawer.Claim(1000);
    myCashdrawer.DeviceEnabled = true;
    myCashdrawer.OpenDrawer();
    myCashdrawer.DeviceEnabled = false;
    myCashdrawer.Release();
    myCashdrawer.Close();
}
```

The code performs the open, claim, device enable, action, device disable, release, and close all in sequence. This means the device is free to be used with other applications. The same solution will be used for the Receipt Printer in a later example when the Receipt Printer is shared between two computers.

3.6.3   Part 3: Build and Test
Let's run the application and test the code.

1. Make sure the cash drawer is connected to the Receipt Printer, the Receipt Printer is connected to the PC, and the OPOS driver has been configured for both devices.

74

2. Make sure the project is set for Debug and build the application.
3. Start the application by hitting F5 or selecting Debug->Start Debugging from the menu.
4. Click on the "Open Cash Drawer" button. The cash drawer should kick open.

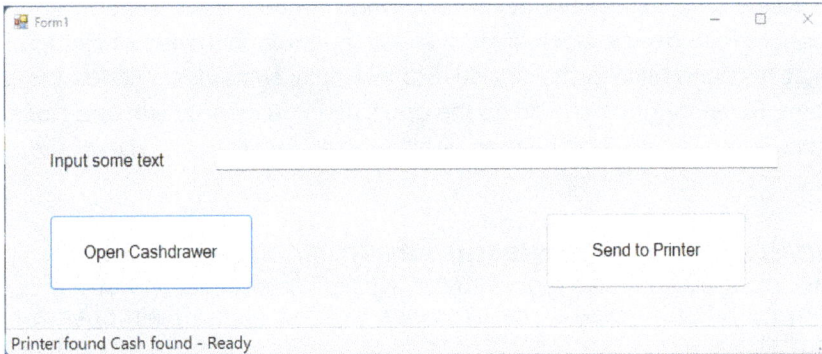

5. Close the cash drawer.
6. Test the print functionality again.
7. Open the cash drawer.
8. You can Close the application.
9. Stop debugging when finished.

3.7   Application Recommendations

The application examples demonstrate the basic functionality of accessing the different POS devices. A real-world application would require a lot more exception checking and architecture design. Here are some recommendations:

1. Try-Catch - Adding Try-Catch around the different PS for .NET method calls will help mitigate application crashes. If a device has a failure during operation, a call to Open or Claim will raise an exception.
2. Uncheck Prefer 32-bit. Windows is only available as a 64-bit OS. There is no need to support a 32-bit build.
3. Logic Device Names – adding the Epson OPOS driver added a bunch of devices available for access. The barcode scanner application will fail to run since there are now multiple "scanner" devices in the system because of the Epson OPOS driver. The application will not know which one to access, and the POS for .NET API will raise an exception. Defining a device with a logical device name and opening the device based on that name will get around the multiple device issue. Write the application with logic device names, and be sure to document those names so technicians in the field know what to set the logic device names to if they are replacing a device in the field.

4. A thread should access the UI thread via an invoke call. Mixing threads with the UI thread will create a sluggish application so do use the "this" keyword when instantiating PosExplorer. The POS .NET DataEvent, DeviceAdded event, and DeviceRemoved event can interact with the UI thread through an invoke call.

5. Keep the claim to the POS device only when needed. If two systems are accessing a single POS device, each system should go through the full UnifiedPOS method to access and release the device. For example, two sales stations accessing the same Receipt Printer should do the open, claim, print, release, and close all in the same method call. The sequence allows each station to access and share the Receipt Printer.

3.8   Summary: From Architecture to Application

The chapter exercises demonstrate POS for .NET's implementations of the UnifiedPOS specification for 5 POS devices. The examples focused on the main operations for the devices, but there are many more methods and properties to explore for each device. The different exercises demonstrated the differences between using OPOS drivers and Service Objects. OPOS driver setup versus Plug and Play was a big difference. Most of the time the devices are going to be attached and the DeviceAdded and DeviceRemove events are not necessary. Different application types, Windows Forms or WPF .NET Framework applications, can be created.

The simulator Service Object had to be removed since it added a second scanner and MSR to the system. The barcode scanner and MSR applications relied on the default device only. When we added the Epson OPOS driver, it added to the number of devices available to the system. Logical device names were used for the Pole Display, Receipt Printer, and cash drawer to zero in the specific device and port. How to handle this situation for Service Objects will be covered in the next chapter on managing Service Objects with POS device manager (POSDM.EXE) and the SOManager utility from Annabooks.

4 POS for .NET Device Management

In the last chapter, the barcode scanner and MSR applications demonstrated how to get an instance of a device using just the device name. The simulator Service Object had to be moved out of the way so a conflict didn't occur. Once the Epson OPOS driver was added, accessing by device name failed since there were multiple devices with the same device name. If the barcode's example Service Object was set as the default or with a logical name, the failure to open the device based on the logical name would not have been a problem. Setting up the Service Object is simple. OPOS drivers are a bit different. A manufacturer has an OPOS driver setup utility for installation at a minimum. As we saw with the Epson OPOS driver utility, logical names were set for the pole display, Receipt Printer, and Cash Drawer device. Going from device to device using different utilities makes the management of POS devices a chore for IT administrators.

POS for .NET offers a couple of Service Object management solutions. One solution is the POSDM.EXE utility, which we used in the previous chapters to list devices. Another solution is to create custom management utilities or scripts, using WMI support. Both methods allow an IT manager to remotely manage POS systems over a network. This chapter will explore these solutions:

- Explore the different features of POSDM.EXE.
- Manage the system using WMI and PowerShell scripts.
- Create a custom Service Object management application in Visual Studio.
- Test application performance.

4.1 POSDM Utility

POSDM.EXE utility was introduced earlier to list the POS devices connected to the system, but the command-line utility provides much more capability. For Service Objects, there are several commands to add device, remove device, set path, set logical name, and get properties. Since OPOS drivers have their setup utility, POSDM only offers to list devices, properties, and logical names for OPOS controls. POSDEM cannot change the properties for an OPOS driver. In addition, POSDM can manage a POS device on a remote machine, which allows for remote management from a central computer. Looking at the help for POSDM yields the following:

Usage:

- POSDM [general switches] <command> [command arguments ...]
- POSDM HELP <command>

General switches:

Switch	Description
/?	Displays this client usage information.
/OUTPUT:<file>	Prints all output to <file> instead of the console.
/MACHINE:<host>	Use <host> as the remote machine. Default is local host.
/USER:<name>	Use <name> as the user account. Prefix with domain separated by a backslash if the user domain is not the current one.
/PASSWORD:<password>	The <password> for the '/USER:<name>' switch.

Commands:

Command	Description
HELP	Prints command-specific help.
ADDDEVICE	Add a physical device (non-Plug and Play).
ADDNAME	Add a name to a device's list of names.
ADDPROPERTY	Add a property to a device.
DELETEDEVICE	Delete a physical device (non-Plug and Play).
DELETENAME	Delete a name from a device's list of names.
DELETEPROPERTY	Delete a property from a device.
DISABLE	Disable an SO on a POS device.
ENABLE	Enable an SO on a POS device.
INFO	Displays the device properties.
LISTDEVICES	List the POS devices on the target <host>.
LISTNAMES	List the names associated with POS devices.
LISTPROPS	List the properties associated with a device.
LISTSOS	List the POS Service Objects on the target <host>.
SETDEFAULT	Set one device as the default of its <type>.
SETPATH	Sets the POS device <path> (non-Plug and Play.)

Command Switches:

Command Switch	Description
/TYPE:<type>	The device type as determined by the SO supplier. e.g., 'Msr' is a type of Magnetic Stripe Reader
/SONAME:<name>	The Service Object name (from the SO supplier). e.g. 'MsrSimulator'
/PATH:<path>	The device's hardware path; e.g., 'COM2'.
/NAME:<name>	A logical name for the device. This is useful in situations where there are multiple instances of the SO, and it also

	provides a means by which an application can refer to an SO without hardcoding which one. Must be unique within <type>.
/INFO	The device's properties are retrieved.

Example Commands:

- List devices on a remote machine called SJJ5:

  POSDM /MACHINE:SJJ5 /USER:"Me User" /PASSWORD:password LISTDEVICES

  A username and password on the remote machine are required.

- Locally set the logical name for the Example Service Object to 'myScanner':

  POSDM ADDNAME myScanner /TYPE:Scanner /SONAME:"Example Scanner"

  The command switches for /TYPE and Service Object name (/SONAME) are needed to specify which scanner is to be set. The command switch /PATH may also be needed if two scanners are using the same Service Object.

- List the local POS devices and the paths to the Service Object DLLs:

  POSDM LISTSOS

  Notice that the physical path does not get listed with this command.

Note: POSDM should be run as Administrator.

4.2 Exercise 4.1 Set Logical Name for the Example Scanner Service Object

A logical name can be used when there are two devices of the same type in the system. In this example, the simulator Service Object will be put back, and the real scanner will be set with a logical name.

1. If you moved the simulator Service Object in the last chapter exercises, move it back to the installed location – "C:\Program Files (x86)\Microsoft Point Of Service\SDK\Samples\Simulator Service Objects".
2. Plug the USB barcode scanner back into the system.
3. Open a command window in administrator mode.
4. Use POSDM to set the logical name for the scanner to "myScanner".

POSDM addname myScanner /TYPE:Scanner "/SONAME:Example Scanner"

5. Use POSDM to check that the name was set correctly:

POSDM LISTNAMES

The resulting output is as follows:

```
Administrator: Command Prompt                                              —  □  ×
    SETPATH                   Sets the POS device <path>. (Non Plug & Play.)

C:\Program Files (x86)\Microsoft Point Of Service>posdm listnames
Type        SoName            Name         Path

Scanner     Example Scanner   myScanner    \\?\HID#VID_0536&PID_0167#6&2BE85541&1&0000#{4D1E55B2-F16F-11CF-88CB-00111
1000030}
LineDisplay DM-D110           LineDisplay  COM4
PosPrinter  TM-T88V           epsonPrinter TM-T88V
CashDrawer  Standard(v1_14_0001) mmfCD     TM-T88V(2)


C:\Program Files (x86)\Microsoft Point Of Service>
```

The LISTNAMES command lists all the devices with their logical names. The simulator service objects don't have assigned logical names, but they are devices that are still available. With a logical name, we can specify in the application the exact device we want to access rather than randomly getting the device that is first in the list. Let's see this in action by modifying the EX31_Bar_Code project.

6. Open Visual Studio and open the EX31_Bar_Code project from the last chapter.
7. Modify the GetDevice line and add the logical name:

```
public Form1()
{
    InitializeComponent();
    _posExplorer = new PosExplorer();
    _posExplorer.DeviceAddedEvent                              +=
_posExplorer_DeviceAddedEvent;
    _posExplorer.DeviceRemovedEvent                            +=
_posExplorer_DeviceRemovedEvent;
    //DeviceInfo device = _posExplorer.GetDevice("Scanner");
    DeviceInfo                        device                   =
_posExplorer.GetDevice("Scanner","myScanner");
```

8. With the barcode scanner connected to the system, run the application. If there are two scanners of the same type in the system, the logical name is used to open a specific scanner.

Even if the device is removed and reinserted, the logical name will persist until you use POSDM to remove the name.

4.3 WMI Support

A batch file can be used to run multiple POSDM commands with a single call. An alternative is to use a PowerShell script with the POS for .NET WMI support. Scripts allow you to create custom commands and take advantage of conditional branching that POSDM, by itself, doesn't offer. When you install the SDK, a WMI namespace called /root/MicrosoftPointOfService is available. This namespace defines four classes, as shown in the following table:

Class	Description
ServiceObject	Represents a POS for .NET Service Object from a management perspective.
PosDevice	Represents the physical device serviced by the Service Object.
LogicalDevice	Represents a logical name assigned to a PosDevice, providing third-party applications with the ability to access a Service Object without conflicting with other applications that may also be accessing the same Service Object.
DeviceProperty	Instances are name/value pairs that can be associated with a PosDevice to store optional configuration data for Service Objects.

The next sections discuss the properties and methods for each of the classes.

4.3.1 The ServiceObject Class

The ServiceObject Class represents the management view of a Service Object.

Property	Access	Key	Description
string Name	Read	X	The name of the Service Object.
string Type	Read	X	The POS Device Class that is implemented by the Service Object.
string UposVersion	Read		The version of the UnifiedPOS standard the Service Object is implementing.
string Path	Read		Path of the Service Object's assembly.
string Version	Read		The version number of the Service Object's assembly.
uint32 Compatibility	Read		The major version of POS for .NET that the Service Object is compatible with.
string Description	Read		A short description of the Service Object.
bool IsPlugNPlay	Read		If TRUE, the Service Object supports Plug and Play.
bool IsLegacy	Read		If TRUE, the device is using a legacy (OPOS) Service Object. If FALSE, the device is using a POS for.NET Service Object.

Method	Description
void AddDevice(string *Path*)	Adds a non-Plug and Play device for this Service Object. *Path* is the hardware path of the non-Plug and Play device to be added.
void DeleteDevice(string *Path*)	Deletes a non-Plug and Play device associated with this Service Object. *Path* is the hardware path of the non-Plug–N-Play device to be deleted.

4.3.2 The PosDevice Class

The PosDevice Class represents a single physical POS device. The class provides properties and methods that are needed to manage that physical device.

Properties	Access	Key	Description
string Type	Read	X	The POS device type or category; for example, scanner, printer, etc.
string SoName	Read	X	The name of the Service Object for this physical device.
string Path	Read	X	The hardware path of a Device. For Plug and Play devices, this path comes from the Plug and Play engine. For non-Plug and Play devices, it is provided via the AddDevice method of ServiceObject. For devices using legacy (OPOS) Service Objects, this may be blank.
string HardwareDescription	Read		The device description of the logical device returned from the registry is used by the Plug and Play engine. This may be blank for devices using legacy (OPOS) Service Objects
bool IsPlugNPlay	Read		If TRUE, the device is a Plug and Play device.
bool IsLegacy	Read		If TRUE, the device is using a legacy (OPOS) Service Object. If FALSE, the device is using a POS for.NET Service Object.
bool Enabled	Read/Write		If TRUE, enable the device. If FALSE, disable the device.
bool Default	Read/Write		If TRUE, the device is the default in a POS Device category.

Method	Description
void AddName(string *Name*)	Adds a logical name for the device. *Name* is the name of the logical device to add. The name must be unique within a device class (type). Logical names are represented by the LogicalDevice class.
void DeleteName(string *Name*)	Deletes the logical name from the device. *Name* is the name of the logical device to delete. Logical names are represented by the LogicalDevice class.
void AddProperty(string *Name*, string *Value*)	Adds a property (a name/value pair) to this device. *Name* is the name of the property. *Value* is the value of the property. Device properties are represented by the Property class.
void DeleteProperty(string *Name*)	Deletes a property from this device. *Name* is the name of the property to be deleted. Device properties are represented by the Property class.
void SetPath(string *Value*)	Sets the physical path for the POS Device. *Value* is the path (i.e. COM1).

4.3.3   LogicalDevice Class

The LogicalDevice class represents a logical device associated with a PosDevice. It provides a naming mechanism so that applications can be developed independently and refer to the same device without conflict.

Property	Access	Key	Description
string Type	Read	X	The POS device category that the logical device belongs to.
string SoName	Read	X	The name of the Service Object.
string Path	Read	X	The path of the physical device.
string Name	Read	X	The name for the logical device.

4.3.4   DeviceProperty Class

The DeviceProperty class represents a name/value pair of a configuration property for a physical device. There may be multiple DeviceProperties associated with a PosDevice.

Property	Access	Key	Description
string Type	Read	X	The POS device category.
string SoName	Read	X	The name of the Service Object.
string Path	Read	X	The path of the physical device.

string Name	Read	X	The name of this property.
string Value	Read		The data of this property.

4.3.5   WMI Code Creator Utility

Working with the WMI can be a bit challenging, but Microsoft developed the WMI Code Creator utility to assist in generating code for C#, VB.NET, and VB scripts. The utility is a free download, which can be used as a guide to writing applications that access the WMI namespaces. With the WMI Code Creator utility, just open the Namespace root\MicrsoftPOintOfService and select the class.

Note: It takes a minute or two for all the namespaces to be listed.

Sample code can be generated for properties.

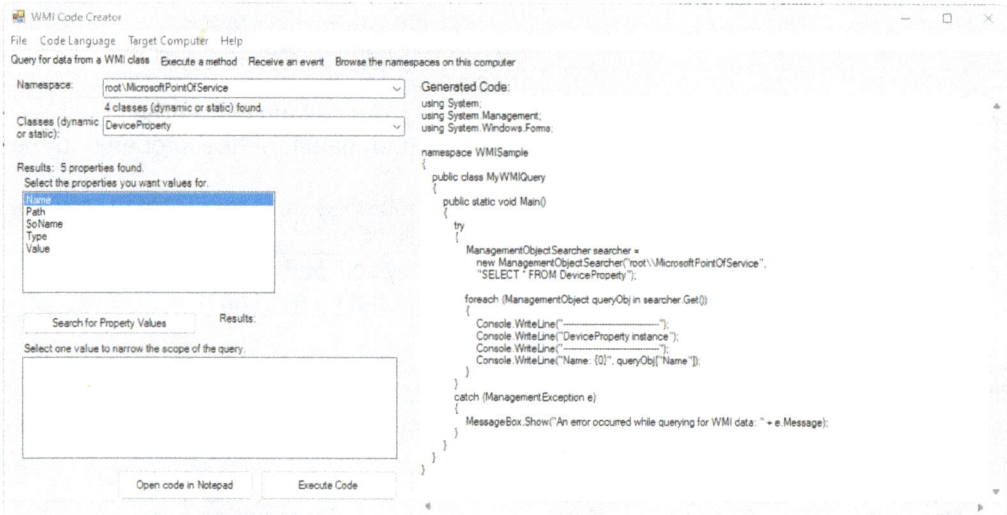

On another tab, code can be generated for methods.

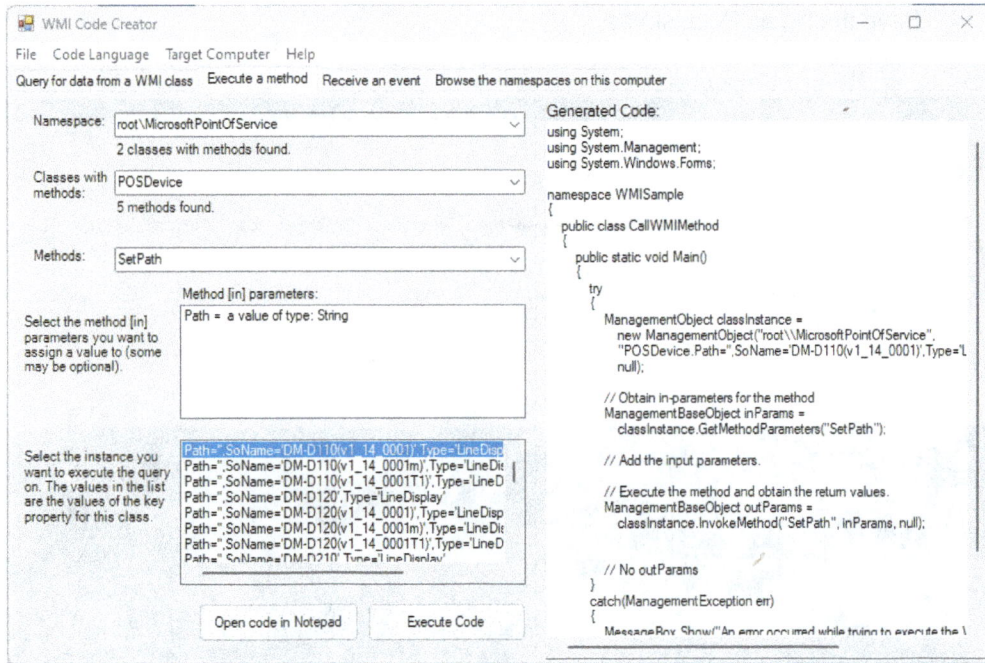

4.4 Exercise 4.2 Using WMI in PowerShell

The following three sub-exercises demonstrate how to use the WMI namespace /root/MicrosoftPointOfService in PowerShell. The WMI Code Creator doesn't output PowerShell script, but it can be used to generate code for VB Script which can be a guide to create the PowerShell script.

4.4.1   Sub Exercise 1: PowerShell Equivalent to POSDM LISTDEVICES

The following steps create a PowerShell script that mimics the POSD LISTDEVICES call.

1. Open PowerShell ISE with administrator privileges.
2. Create a new script.
3. Enter the following:

```
$POSDeviceList = get-wmiobject -class "PosDevice" -namespace
"root\MicrosoftPointOfService"
write-host "List of Devices"
foreach($POSDevice in $POSDeviceList){
    write-host "Type:    " $POSDevice.Type
    write-host "SoName: " $POSDevice.SoName
    write-host "Path:    " $POSDevice.Path
    write-host ""
}
```

4. Save the file as PosList.vbs.
5. Run the script in PowerShell ISE and you will get a list of devices.

4.4.2 Sub Exercise 2: PowerShell Equivalent to POSDM LISTNAMES
The following steps create a PowerShell script that mimics the POSD LISTNAMES call.

1. Open PowerShell ISE with administrator privileges.
2. Create a new script.
3. Enter the following:

```
$POSNameList = get-wmiobject -class "LogicalDevice" -namespace
"root\MicrosoftPointOfService"
write-host "List of Logical Names"
foreach($POSName in $POSNameList){
    write-host "Type:     " $POSName.Type
    write-host "SoName: " $POSName.SoName
    write-host "Logical Name: " $POSName.Name
    write-host "Path:     " $POSName.Path
    write-host ""
}
```

4. Save the file as PosListNames.ps1.

5. Run the script in PowerShell ISE and you will get a list of logical device names. Make sure the barcode scanner is plugged in so you can see the myScanner name.

```
#Copyright (c) 2006-2024 Annabooks, All Rights Reserved
#PS1 POSDM LISTNAMES script
$POSNameList = get-wmiobject -class "LogicalDevice" -namespace "root\MicrosoftPointOfService"
write-host "List of Logical Names"
foreach($POSName in $POSNameList){
    write-host "Type:     " $POSName.Type
    write-host "SoName:   " $POSName.SoName
    write-host "Logical Name: " $POSName.Name
    write-host "Path:     " $POSName.Path
    write-host ""
}
```

```
PS C:\WINDOWS\system32> C:\NETPOS\PosListNames.ps1
List of Logical Names
Type:     LineDisplay
SoName:   DM-D110
Logical Name:  LineDisplay
Path:     COM4

Type:     PosPrinter
SoName:   TM-T88V
Logical Name:  epsonPrinter
Path:     TM-T88V

Type:     CashDrawer
SoName:   Standard(v1_14_0001)
Logical Name:  mmfCD
Path:     TM-T88V(2)
```

4.4.3 Sub Exercise 3: Custom Information Using the ServiceObject Class
Here is a script that gets different information from the ServiceObject class.

1. Open PowerShell ISE with administrator privileges.
2. Create a new script.
3. Enter the following:

```
$POSSOList = get-wmiobject -class "ServiceObject" -namespace
"root\MicrosoftPointOfService"
write-host "List of Devices"
foreach($POSSO in $POSSOList){
    write-host "Type:     " $POSSO.Type
    write-host "SoName: " $POSSO.Name
    write-host "UPOS ver: " $POSSO.UposVersion
    write-host "Description:     " $POSSO.Description
    write-host ""
}
```

4. Save the file as PosSOprops.ps1.
5. Run the script in PowerShell ISE.

```
    1    #Copyright (c) 2006-2024 Annabooks, All Rights Reserved
    2    #PS1 Service Objects Information script
    3    $POSSOList = get-wmiobject -class "ServiceObject" -namespace "root\MicrosoftPointOfService"
    4    write-host "List of Devices"
    5    foreach($POSSO in $POSSOList){
    6        write-host "Type:      " $POSSO.Type
    7        write-host "SoName: " $POSSO.Name
    8        write-host "UPOS ver: " $POSSO.UposVersion
    9        write-host "Description:   " $POSSO.Description
   10        write-host ""
   11    }
```

```
PS C:\WINDOWS\system32> C:\NETPOS\PosSOprops.ps1
List of Devices
Type:      CashDrawer
SoName:  Microsoft CashDrawer Simulator
UPOS ver:  1.14
Description:    Microsoft Cash Drawer Simulator

Type:      CheckScanner
SoName:  Microsoft CheckScanner Simulator
UPOS ver:  1.14
Description:    Microsoft Check Scanner Simulator

Type:      Keylock
SoName:  Microsoft Keylock Simulator
UPOS ver:  1.14
Description:    Service object for Microsoft Keylock Simulator
```

4.5   Exercise 4.3: Custom Management Solution

Scripts can be created to test the WMI API or create a quick batch solution. You might want to provide customers or field technicians with a management / setup utility that has a more user-friendly interface. The WMI calls can be performed in a C# or VB.NET application. The WMI Code Creator utility can generate code for either language. Calls to the PosDevice, ServiceObject, LogicalDevice, and DeviceProperty classes can be made from within a .NET Framework application. The application can support local and remote POS computers. The application in this exercise will simply list all of the devices in the system.

The application is going to list some of the properties for each POS device in the system: SoName, Type, Enabled, and Path. The WMI Code Creator can generate sample code to get the values for these properties. You have to click on each property to get the code sample and combine the results.

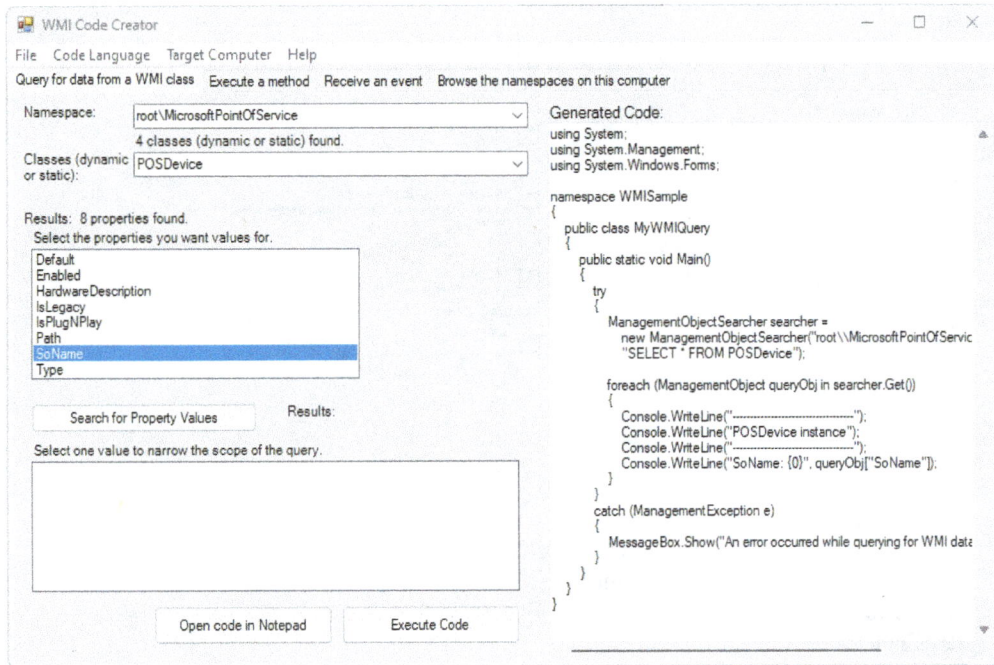

4.5.1 Part 1: Create the Project

With the WMI Code Creator information in hand, let's create the project.

12. Open Visual Studio.
13. Click on "Create a new Project".
14. Set the language to C# and search for the "Windows Forms App (.NET Framework)" template.

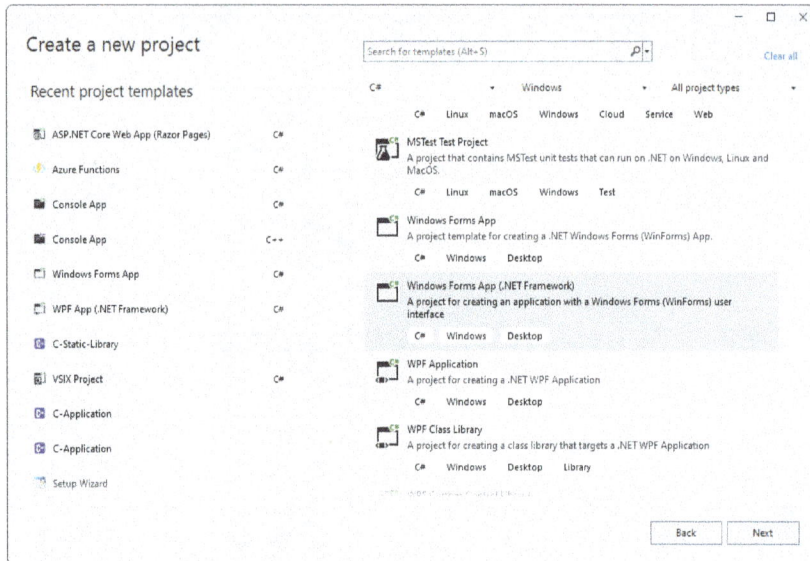

15. Click Next.
16. Name the project EX43_POSDEVICES.
17. Set the location for the project.
18. Keep the Framework set to .NET Framework 4.8.
19. Click Create.
20. Adjust the size of Form1.
21. On the form, add a ListBox control, button control, and a label with the following properties:

> ListBox:
>> Name: lstPOSDevices
>> HorizontalScrollbar: True

> Button:
>> Name: btnGetPOSDevices
>> Text: Get POS Devices
>> Font: Arial, 12pt

> Label
>> Name: lblPOSDevices
>> Text: POS Devices:
>> Font: Arial, Regular 12pt

22. Save the project.

4.5.2   Part 2: Add the Code Behind the Button

1. Open Form1.cs in code view.
2. At the top of the code, before the Form1 class, add the imports for System.Management:

```
using System;
using System.Collections.Generic;
using System.ComponentModel;
using System.Data;
using System.Drawing;
using System.Linq;
using System.Text;
using System.Threading.Tasks;
using System.Windows.Forms;
using System.Management;
```

3. Go to Form1.cs design view.
4. Double-click on the button to add the button handler code.
5. Add the following code to the btnGetPOSDevices_Click method:

```
private void btnGetPOSDevices_Click(object sender, EventArgs e)
{
    String POSDType;
    String POSSoName;
    String finalString;
    String enabled = "N";


    ManagementObjectSearcher          mySearch          =          new
ManagementObjectSearcher("root\\MicrosoftPointOfService",
"SELECT * FROM POSDevice");
    foreach (ManagementObject device in mySearch.Get())
    {

        enabled = "N";
        if          (String.Equals(device["Enabled"].ToString(),
"True"));
        {
            enabled = "Y";
        }

        POSDType = device["Type"].ToString();
        POSSoName = device["SoName"].ToString();
```

```
        finalString = POSDType.PadRight(15) + (char)Keys.Tab +
POSSoName.PadRight(40)   +   (char)Keys.Tab   +   enabled   +
(char)Keys.Tab + device["Path"];
```

```
        lstPOSDevices.Items.Add(finalString);
```

```
    }
}
```

The code is a variation on what is generated by WMI Code Creator. The code performs a query on the MicrosoftPointOfService root for the POSDevice namespace. The foreach makes the Get() method call to get each device in the list. The different properties are then put into a string that is added to the list box.

6. Save the project.

4.5.3   Part 3: Build and Test
With the code created, let's test the application.

1. Make sure that the project is set for Debug, and build the application.
2. Change the project properties and uncheck Prefer 32-bit from the Build options.

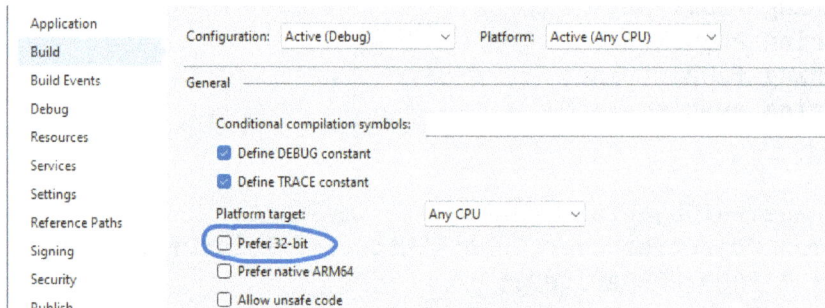

3. Start the application by hitting F5 or selecting Debug->Start Debugging from the menu.
4. Click the Get POS Devices button, and the POS devices for the system should be listed, which is similar to running posdm.exe listdevices.

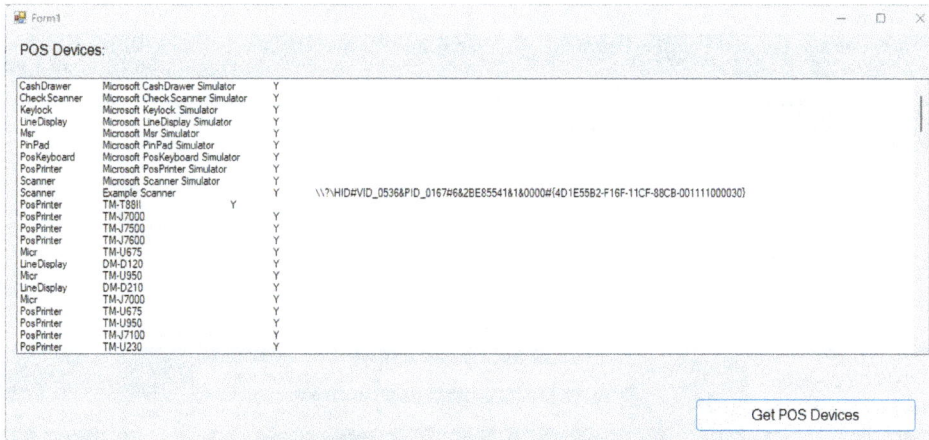

An application could be created to include methods that set logical names and device paths for a Service Object or simply enable or disable a Service Object.

Note: If you plan to create a utility that is going to perform the method calls, the application will have to run as administrator.

4.6   Service Object Manager (SOManager.exe)

The WMI extensions offer different capabilities to manage Service Objects. From the last example, you could create a GUI version of POSDM, and someone already has. Several years ago, I created a GUI version of POSDM called the Service Object Manager utility, which is a free download from Annabooks.com. The utility uses the WMI extensions and provides most (not all) of the same functionality as POSDM. The current version supports POS for .NET v1.14.1.

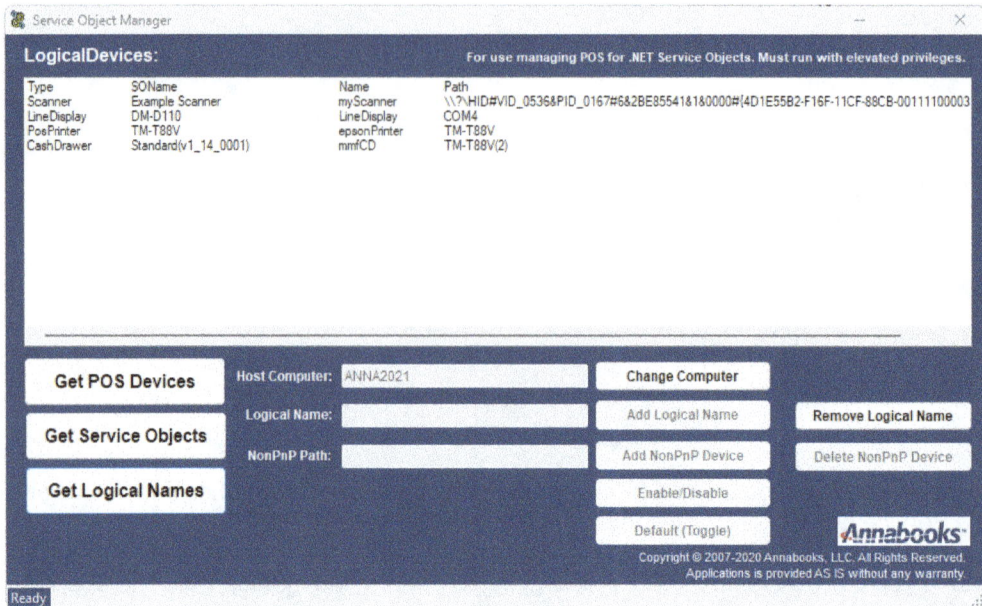

4.7 Big Warning! Windows Updates Breaking POS for .NET WMI

The issue is important enough to warrant its own section. You might be working on a project and the applications and WMI are working properly. Then, you stop for a few months and attempt to restart the project again. This time nothing seems to work. The simple solution:

Reinstall the POS for .NET 1.14.1 SDK and Runtime.

This has happened several times and to some developers who have reached out. Windows updates appear to break the POS for .NET WMI. My guess is that there is not much visibility of POS for .NET at Microsoft and that it is not part of the test cycle. A simple re-install appears to fix any issues, and you can continue with your development. Please, keep this in mind during your development or when questions arise from the field.

4.8 POS Performance Monitoring – Walk-Through Example

The last chapter covered the basic functionality of accessing devices, and this chapter focused on how to manage Service Objects. There will be a point, whether in development or the field, that the performance of the application has to be tested. The POS devices can be monitored using Perfmon. You can track good and bad scans to see if a device is about to fail or is not being used properly. Performance monitoring comes free with the Base Service Objects. You will have to implement support for SO using the Basic inheritance. The POS performance counters are a service in Windows. In this walk-through example, the barcode scanner and the EX31_Bar_Code project will be used to demonstrate performance monitoring.

1. By default, the Point of Service Performance Counters service is disabled. You need to enable and start the Point of Service Performance Counters in the Administrator Tools->Services as shown in the picture below.

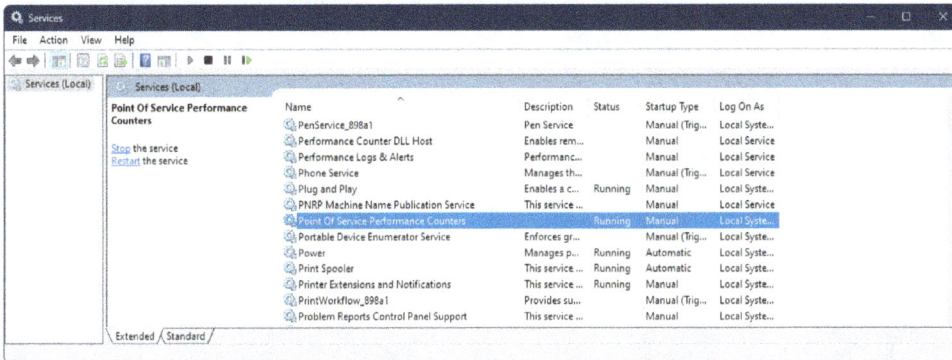

2. Plug in the barcode scanner.
3. Once the service has been enabled, you will be able to add tracing in Perfmon for POS devices like the barcode scanner, MSR, cash drawer, etc. Start Performance Monitor.
4. Click on the (+) plus symbol and search for the Example Scanner.
5. Click on Example Scanner.
6. Click Add.

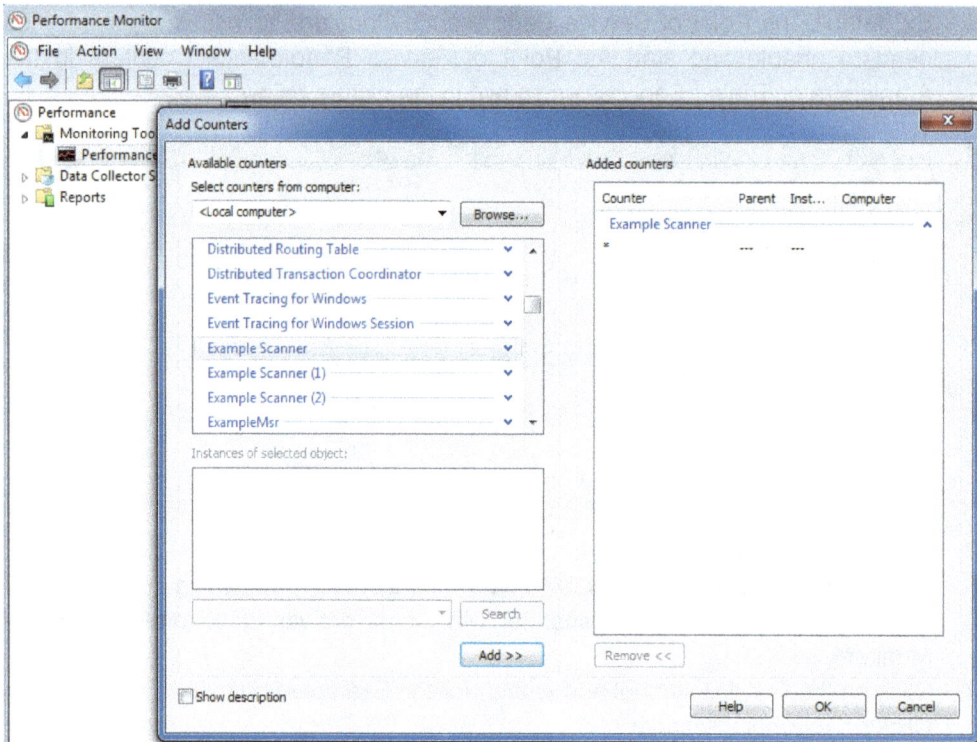

7. Click OK. Three counters will be added to the list.
8. Run either the test application or the application developed in the previous chapter. Click on a few barcodes, and you can see the performance monitor tracking the device.

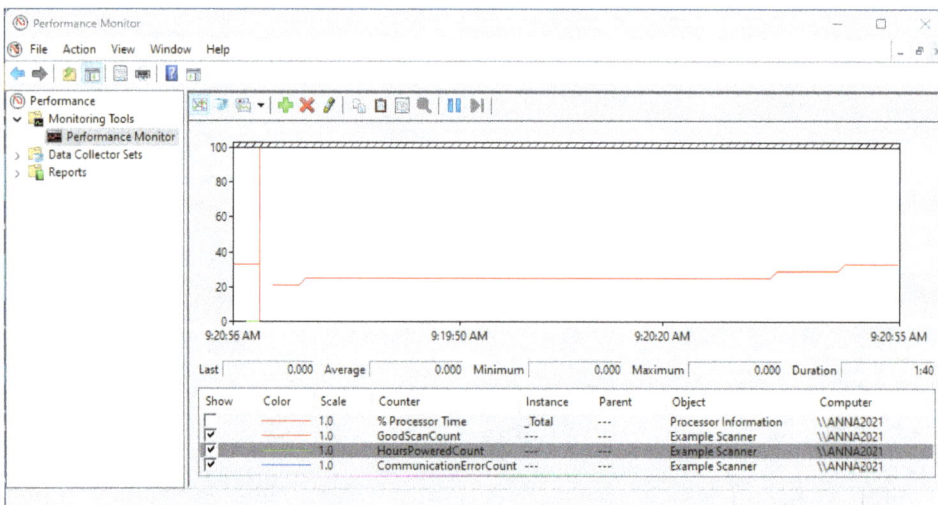

9. Close the Performance Monitor and disable the service when finished.

4.9 Summary: Take Control

OPOS drivers require setup utilities from the manufactures. This scheme forces administrators to manage several individual programs on each local machine. POS for .NET breaks the paradigm with Service Objects. POS for .NET provides a command-line utility and WMI extensions to manage all Service Objects, either locally or remotely.

POSDM.EXE contains all the basic functionality for managing a system, either locally or remotely via a command-line interface. The WMI extensions allow you to create custom Service Object setup utilities or scripts. The ability to create custom Service Object management tools addresses the needs of different users:

- Administrators who want to centralize management and control of their POS systems in their network.
- Assists POS Application developers with setting up logical names for use with the application and tool to automate the application setup.
- Provides customers with a debug tool to troubleshoot SO configuration issues.
- Assists POS device manufacturers who want to deliver a server object setup with their platforms.

We now know how to manage Service Objects and write POS for .NET applications that call for them. The next chapter looks at how to create Service Objects.

5 POS for .NET Service Objects

Finding Service Objects and OPOS drivers is a major step and part of the system architecture discussion in Chapter 2. The last chapter discussed the advantages of Service Objects over OPOS drivers when it comes to manageability. Even though POS for .NET has been around for some time, some POS device manufacturers have opted to stick with OPOS drivers. The cost-benefit of switching from an OPOS driver to a Service Object is probably low value. As we will see in the next couple of chapters, .NET Core programming offers a different path.

There are some devices that are very simple, and Service Objects could easily be created for them. Receipt printers are too complex, and the support should come from the POS device manufacture. There are others like a cash drawer, scale, or anything with a serial port and simple serial protocol that are simple enough to create Service Objects for. The example Service Objects that come with the SDK are a good starting point. This chapter will cover the following:

- A High-level overview of the classes to create a Service Object.
- A walk-through of creating a Service Object for a weight scale.
- Modifying the Example Service Object to hardcode a PnP ID.

5.1 Simplifying Service Object Development with Interface, Basic, and Base Classes

There are 36 POS devices called out in the UnifiedPOS specification, and all 36 are supported in POS for .NET. There are different levels of support for each device in POS for .NET. Three different classes make up the support.

```
┌─────────────────────────────┐
│         POS Common          │
└─────────────────────────────┘
              │
              ▼
┌─────────────────────────────┐
│     36 Interface Classes    │
└─────────────────────────────┘
              │
              ▼
┌─────────────────────────────┐
│      36 Basic Classes       │
└─────────────────────────────┘
              │
              ▼
┌─────────────────────────────┐
│        9 Base Class         │
└─────────────────────────────┘
```

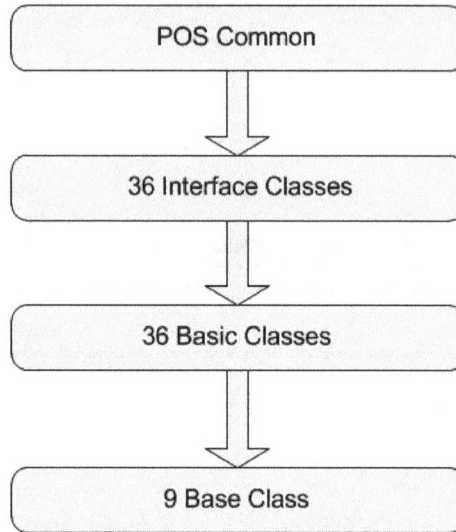

5.1.1   Interface Class

The POS for .NET namespace architecture allows for additional POS devices to be added to the list, as well as changes to the UnifiedPOS specification. The namespaces for the Service Object class are derived from PosCommon. All 36 hardware devices have an abstract Interface Class derived from PosCommon, such as the CashDrawer class. The Interface Class addresses the entry points called out in the UnifiedPOS specification. Very minimal functionality is provided. One would not use the Interface Class to create a Service Object.

5.1.2   Basic Class

Basic Classes contain basic functional support for all 36 devices. POS for .NET provides generic support for: opening, claiming, and enabling the device; device statistics; and management of event delivery to the application. The Basic Classes are used to create the Service Object, and there are 36 of them:

- BeltBasic
- BiometricsBasic
- Bill AcceptorBasic
- BillDispenserBasic
- BumpBarBasic
- CashChangerBasic
- CashDrawerBasic
- CatBasic
- CheckScannerBasic
- CoinAcceptorBasic
- CoinDispenserBasic
- ElectronicJournalBasic
- ElectronicValueRWBasic

- FiscalPrinterBasic
- GateBasic
- HardTotalsBasic
- ImageScannerBasic
- ItemDispenserBasic
- KeylockBasic
- LightsBasic
- LineDisplayBasic
- MicrBasic
- MotionSensorBasic
- MsrBasic
- PinPadBasic
- PointCardRWBasic

100

- PosKeyboardBasic
- PosPowerBasic
- PosPrinterBasic
- RemoteOrderDisplayBasic
- RFIDScannerBasic

- ScaleBasic
- ScannerBasic
- SignatureCaptureBasic
- SmartCardRWBasic
- ToneIndicatorBasic

5.1.3 Base Classes

To make Service Object development even easier, POS for .NET comes with a set of 9 Base Classes that are fully functional implementations. The 9 Base Classes are for the most common POS devices:

- CashDrawerBase
- CheckScannerBase
- LineDisplayBase
- MsrBase
- PinPadBase
- PosKeyboardBase
- PosPrinterBase
- RFIDScannerBase
- ScannerBase

These 9 classes derive support from their counterpart Basic Classes. The device classes share similar class members, like Open method, Close method, CheckHealth, DirectIO, and various properties, which are inherited from the higher-layer classes. There are class members that are unique for each device, like OpenDrawerImpl for the CashDrawerBase Class. You can also override the base methods and properties for custom processing. The POS for .NET help file found in the SDK details the class members for each device.

The Base Classes provide a more complete implementation that can speed up the development process. The 9 devices were chosen because of their popularity in the POS market. The Cash Drawer example in this chapter uses the Base Class. Please see the POS for .NET help for more details about all of the class members for each Base and Basic Class.

When writing a Service Object, start with the Base Class if your device is one of the 9; otherwise, you will use the inheritance from a Basic Class.

5.2 POS Scale Service Object Background

The custom POS Scale Service Object will be created from the ScaleBasic Class. In my search to find devices to write about Service Objects, I came across the Avery-Berkel 6710 scale. The scale has a serial interface and a programmer's guide, which makes it a great candidate for creating a custom Service Object. As a bonus, an OPOS driver was available to compare the final Service Object against.

Understanding communication protocols is the real trick to writing the Service Object. From the manual, the commands are single-string commands followed by a carriage return. Each command expects a response back in a specific format. For example, sending a 'W' to get the weight returns the data in the following format:

<LF>xx.xxxUU<CR><LF>Shh<CR><ETX>

The actual weight data is the xx.xxxUU, which are the weight (xx.xxx) and the unit of measurement (UU ie. UU=LB for pounds). Shh is the status of the scale. If there was an error, the following is returned:

<LF>Shh<CR><ETX>

Since an error can be returned instead of the weight, you need to check the error status before you can accept the weight value. When writing a Service Object for others to use, documenting the error conditions is important. The application programmer needs to account for error conditions and to choose the appropriate response.

5.3 Exercise 5.1: POS Scale Test Application

The first step was to create a test application that performs the basic commands to get weight (W), zero scale (Z), and check status(S). The W command returns 15 characters to the serial port. S and Z return 6 characters. The For-loops read each character in, so we can set up the proper format for error status checks and displaying the weight. Using arrays helps to pick out and combine the characters needed.

Note: The returned scale status bytes didn't match the documentation. The application was used to check the error codes returned from the scale firmware.

5.3.1   Part 1: Create the Application and Form

The application will have three buttons that represent the three commands to send to the scale. A list box will show the result.

> 12. Open Visual Studio.
> 13. Click on "Create a new Project".
> 14. Set the language to C# and search for "Windows Forms App (.NET Framework)" template.

102

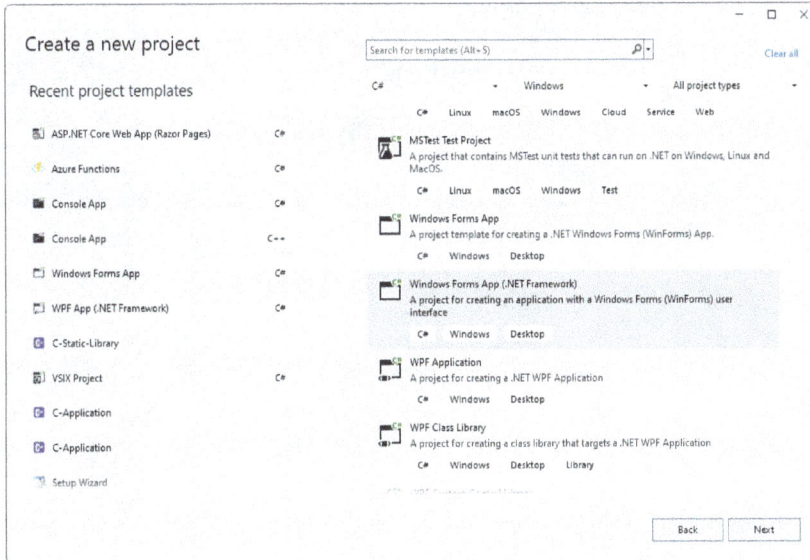

15. Click Next.
16. Name the project EX51_Scale_Test.
17. Set the location for the project.
18. Keep the Framework set to .NET Framework 4.8.
19. Click Create.
20. Adjust the size of Form1 to accommodate long numbers.
21. On the form, add the following controls and the properties:

    ListBox:
        Name: lstResults

    Button:
        Name: btnGetWeight
        Text: Get Weight
        Font: Arial, Regular, 12pt

    Button:
        Name: btnZeroSCale
        Text: Zero Scale
        Font: Arial, Regular, 12pt

    Button:
        Name: btnGetScaleStatus
        Text: Scale Status
        Font: Arial, Regular, 12pt

    StatusStrip:
        Click the drop-down to add: StatusLabel

103

Name: TTStatus
Text: Ready
Font: Arial, Regular, 12pt

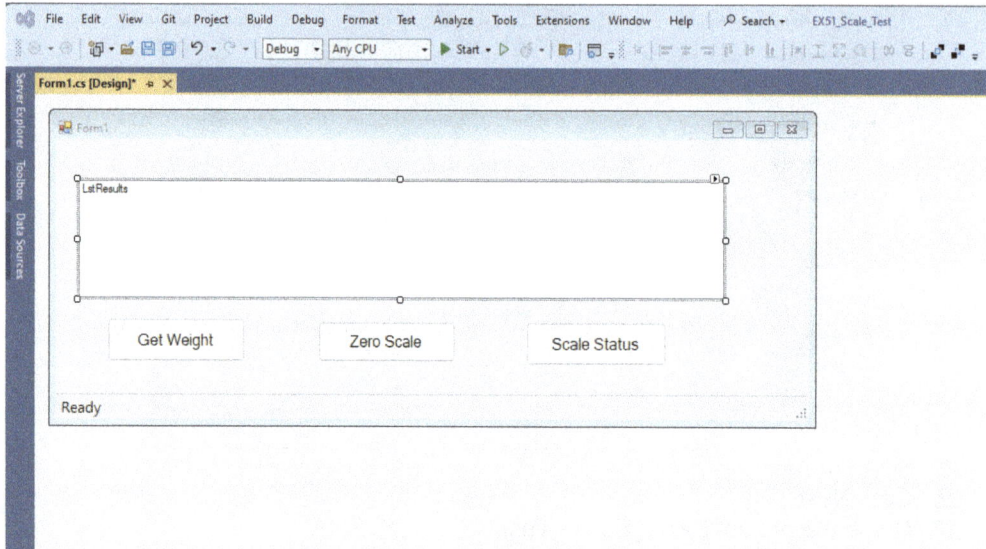

22. Save the project.

5.3.2   Part 2: Adding the Code

With the UI created, the next step is to add the code.

1. Open Form1.cs in code view.
2. Add the serial.IO.ports and system.threading includes:

```
using System;
using System.Collections.Generic;
using System.ComponentModel;
using System.Data;
using System.Drawing;
using System.Linq;
using System.Text;
using System.Threading.Tasks;
using System.Windows.Forms;
using System.Threading;
using System.IO.Ports;
```

3. The scale is connected to the computer via serial port. The next step is to create an instance of the serial port. Before the Form1() method add the following:

```
namespace EX51_Scale_Test
{
    public partial class Form1 : Form
    {

        private SerialPort sPort = new SerialPort("COM4", 9600,
Parity.Even, 7, StopBits.One);

        public Form1()
```

4. Go back to viewing Form1.cs in Designer, and double-click on each of the buttons to add the button handler code to the Form1.cs.
5. Add the following code to the btnGetWeight_Click handler method:

```
private void btnGetWeight_Click(object sender, EventArgs e)
{

    Char[] theWeight = new char[16];
    sPort.Open();
    sPort.Write("W" + (char)13);
    Thread.Sleep(1000);

    try
    {
        theWeight                                              =
sPort.ReadTo(((char)3).ToString()).ToArray(); //Read to the end
of the line ETX
    }
    catch (Exception ex) {

        TTStatus.Text = "Error Reading Weight";

    }
    //Based on a successful result

    if ((theWeight[1].ToString() + theWeight[2].ToString() +
theWeight[3].ToString()) == "S10")
    {
        TTStatus.Text = "Scale in motion";
```

```
    }
    else if ((theWeight[1].ToString() + theWeight[2].ToString()
+ theWeight[3].ToString()) == "S11")
    {
        TTStatus.Text = "Scale Over Capacity";
    }
    else
    {
        //Display the weight
        lstResults.Items.Add(theWeight[1].ToString()          +
theWeight[2].ToString()      +       theWeight[3].ToString()      +
theWeight[4].ToString()      +       theWeight[5].ToString()      +
theWeight[6].ToString()      +       theWeight[7].ToString()      +
theWeight[8].ToString());
    }

    sPort.Close();
}
```

6. Add the following code to the btnZeroScale_Click handler method:

```
private void btnZeroScale_Click(object sender, EventArgs e)
{

    Char[] theZeroStatus = new char[6];
    sPort.Open();
    sPort.Write("Z" + (char)13);
    Thread.Sleep(1000);

    try
    {
        theZeroStatus                                          =
sPort.ReadTo(((char)3).ToString()).ToArray();
    }
    catch (Exception ex)
    {
        TTStatus.Text = "Error with Zeroing scale";
    }

    //Communicate the result
```

```
    if              ((theZeroStatus[1].ToString()                 +
theZeroStatus[2].ToString()  +  theZeroStatus[3].ToString())  ==
"S10")
    {
        TTStatus.Text = "Scale in motion";
    }
    else        if          ((theZeroStatus[1].ToString()        +
theZeroStatus[2].ToString()  +  theZeroStatus[3].ToString())  ==
"S20")
    {
        TTStatus.Text = "Scale at Zero";
    }
    else
    {
        TTStatus.Text = "Unknown Zero Scale Result";
    }

    sPort.Close();

}
```

7. Add the following code to the btnGetScaleStatus_Click handler method:

```
private void btnGetScaleStatus_Click(object sender, EventArgs e)
{

    Char[] theStatus = new char[6];
    sPort.Open();
    sPort.Write("S" + (char)13);
    Thread.Sleep(1000);

    try
    {
        theStatus                                               =
sPort.ReadTo(((char)3).ToString()).ToArray();
    }
    catch (Exception ex)
    {
        TTStatus.Text = "Error Reading State";
    }
```

```
    if  ((theStatus[1].ToString()  +  theStatus[2].ToString()  +
theStatus[3].ToString()) == "S10")
    {
        TTStatus.Text = "Scale in motion";
    }
    else if ((theStatus[1].ToString() + theStatus[2].ToString()
+ theStatus[3].ToString()) == "S20")
    {
        TTStatus.Text = "Scale at Zero and Stable";
    }
    else if ((theStatus[1].ToString() + theStatus[2].ToString()
+ theStatus[3].ToString()) == "S00")
    {
        TTStatus.Text = "Scale Not at Zero and Stable";
    }
    else
    {
        TTStatus.Text = "Unknown Scale result";
    }

    sPort.Close();

}
```

8. Save the project

5.3.3 Part 3: Build and Test
With the code added, let's test the application.

1. Make sure the scale is connected to the computer via the serial port or USB-to-serial adapter.
2. Double-check that the COM port in Device Manager matches the COM port used in the code.
3. Change the project properties and uncheck Prefer 32-bit from the Build options.

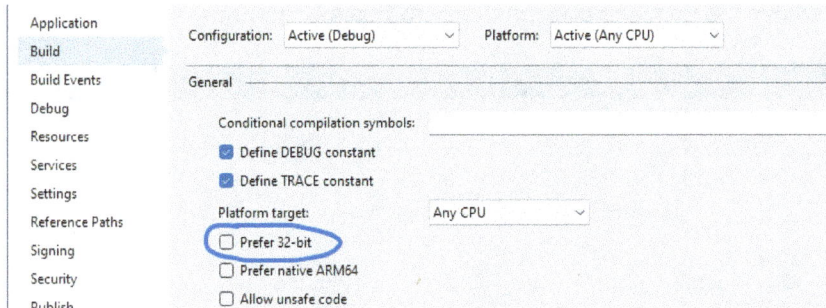

4. Start the application by hitting F5 or selecting Debug->Start Debugging from the menu.
5. Put something on the scale and click on the "Get Weight" button. The weight should show up in the list box.

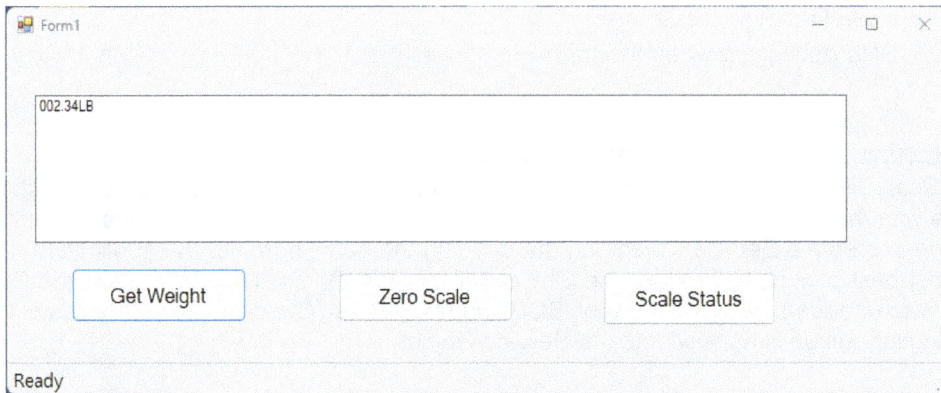

6. Click the "Scale Status" button. The status will show that with something on the scale, the scale is stable and not at zero

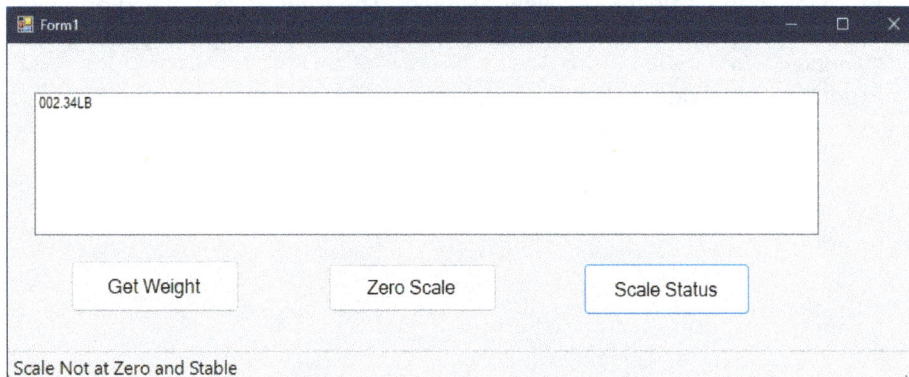

109

7. Remove the item and click on the Zero Scale button.

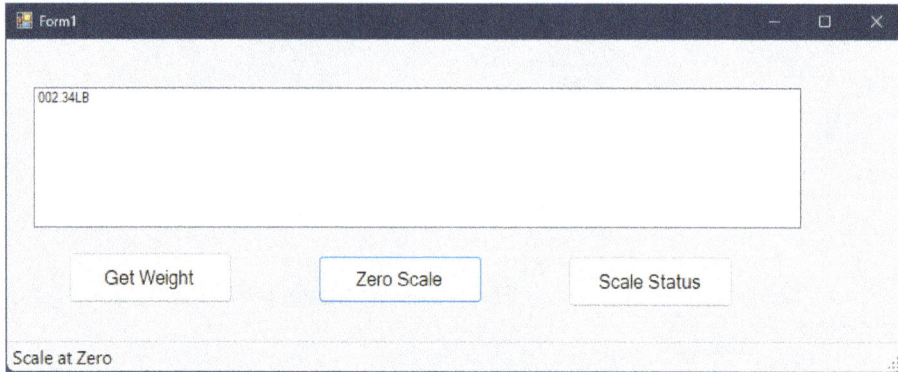

8. You can also gently tap the scale and click on the "Scale Status" button to get a scale-in-motion result.
9. Stop debugging when finished.

5.4 Exercise 5.2: POS Scale Service Object

The Scale Test application in the last exercises proved that we can communicate with the scale with the basic commands. You could just put the code into your application without having to create a Service Object, but hardcoding the serial port into the application goes against best practices. The Service Object will provide the flexibility to change the COM port when needed using POSDM or SOManager utilities. The code from the Scale Test application will be integrated into the Service Object.

5.4.1 Part 1: Project Setup

The first step is to set up the project.

1. Open Visual Studio.
2. From the menu, select File->New->Project. The New Project dialog appears.
3. We want to create a new C# Class library. From the Project Types, select C# / Windows.
4. From the Templates, select Class Library (.NET Framework).

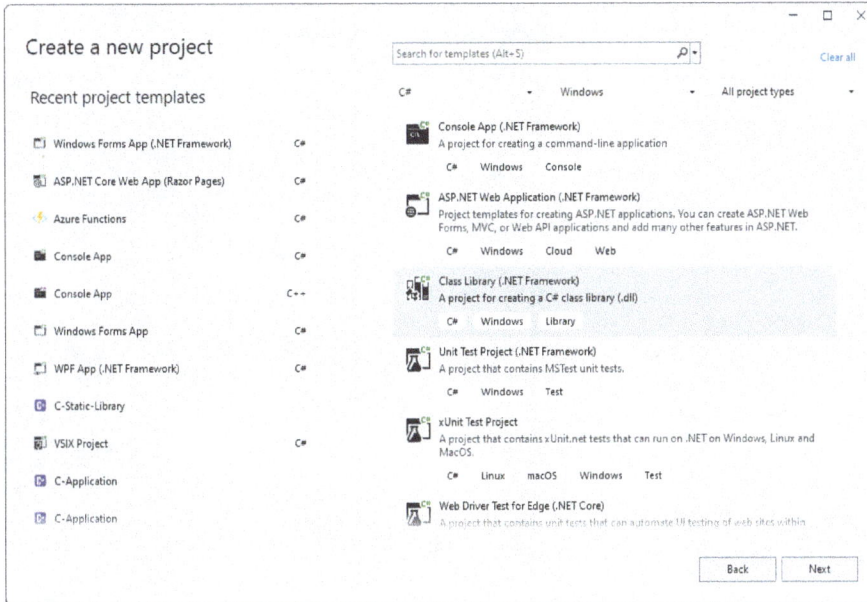

5. Name the project AveryBerkel6710SO.
6. Click Create.
7. The next step is to add the two POS for .NET resources to the project. First, we will add the POS for .NET components. From the menu, select Project->Add Reference. This will open the Add Reference dialog.
8. Click on the Browse tab and locate the Microsoft.PointOfService.ControlBase.dll found under C:\Program Files(x86)\Microsoft Point of Service\SDK. This library is used for creating Service Objects.
9. Click on the OK button.
10. Repeat steps 7 through 9, and add the reference to Microsoft.PointOfService.dll.

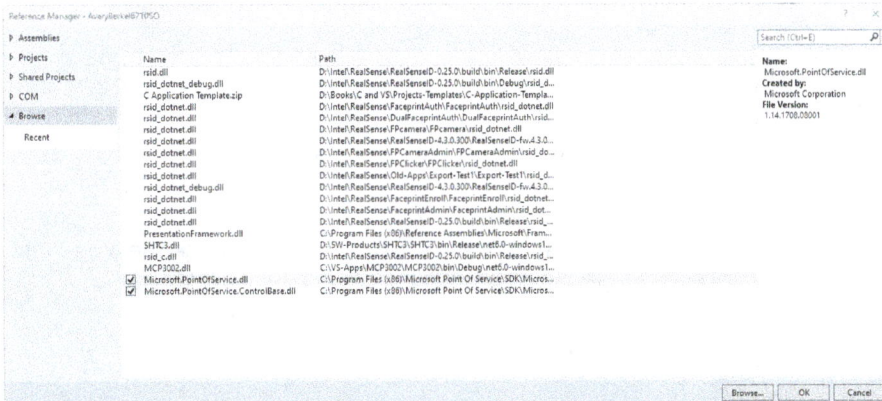

Both references will appear in Solution Explorer.

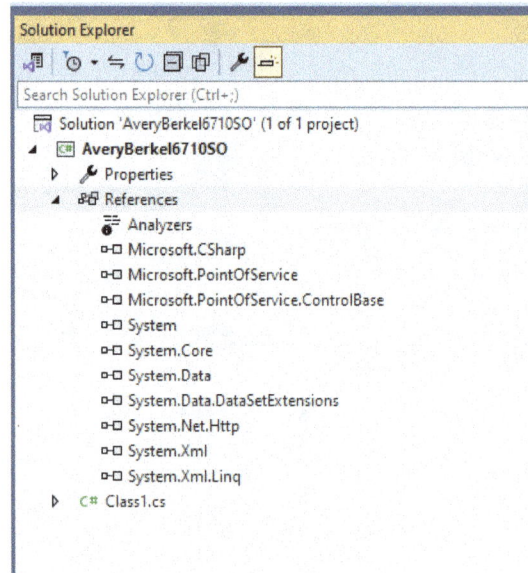

11. The next step is to properly name the class. In the Solution Explorer pane, click on the Class1.cs file.
12. In the properties box, change the file name to AveryBerkel6710SO.cs.

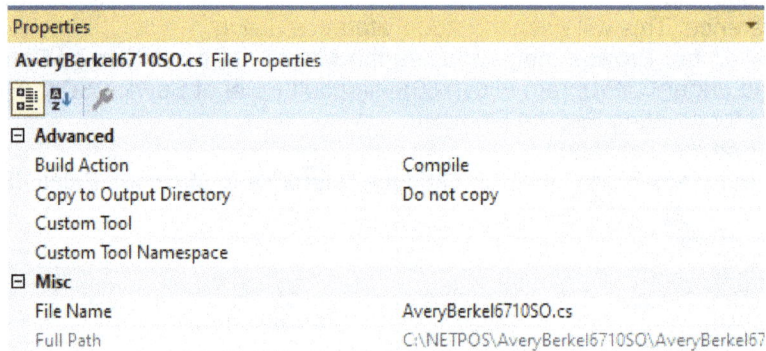

13. Hit enter.
14. You will be prompted to change the name of the class. Click the Yes button, and Visual Studio will change the class name to AveryBerkel6710SO.

Microsoft Visual Studio

? You are renaming a file. Would you also like to perform a rename in this project of all references to the code element 'Class1'?

Yes   No

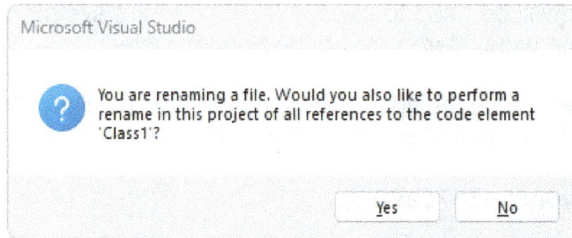

15. Finally, add the imports needed to the top of the project.

```
using System;
using System.Collections.Generic;
using System.Linq;
using System.Text;
using System.Threading.Tasks;
using System.IO.Ports;
using Microsoft.PointOfService;
using Microsoft.PointOfService.BasicServiceObjects;
using System.Threading;
```

16. Save the project.

5.4.2   Part 2: Filling in the POS Scale Service Object
Now, we will add the specific code for the Scale SO.

1. First, we will want to add the Service Object attribute entry point for the DLL. In the AveryBerkel6710SO.cs code view, place the cursor before the AveryBerkel6710SO class and add the following:

```
namespace AveryBerkel6710SO
{
    [ServiceObject(DeviceType.Scale, "Avery Berkel 6710 Scale",
"Annabooks AB 6710 Scale SO", 1, 14)]

    public class AveryBerkel6710SO
```

The entry point defines the following:

      A. Device Type: **Scale**
      B. Service Object Name: **Avery Berkel 6710 Scale**
      C. Description: **SJJ EMS AB 6710 Scale SO**
      D. Major Version: **1**
      E. Minor Version: **14**

113

The entry information will be used to expose the DLL to the system. POSDM or Service Object Manager will be used later to set up the Service Object.

2. Second, we will make the class inherit the ScaleBasic class. Place the cursor next to the public class AveryBerkel6710SO, and enter the following: ": Scale".

```
namespace AveryBerkel6710SO
{
    [ServiceObject(DeviceType.Scale, "Avery Berkel 6710 Scale",
"Annabooks AB 6710 Scale SO", 1, 14)]
    public class AveryBerkel6710SO : Scale
    {

    }
}
```

3. A red squiggly underline appears under the class name, which signals there is some missing code in the library. A little yellow symbol also appears. Click the drop-down

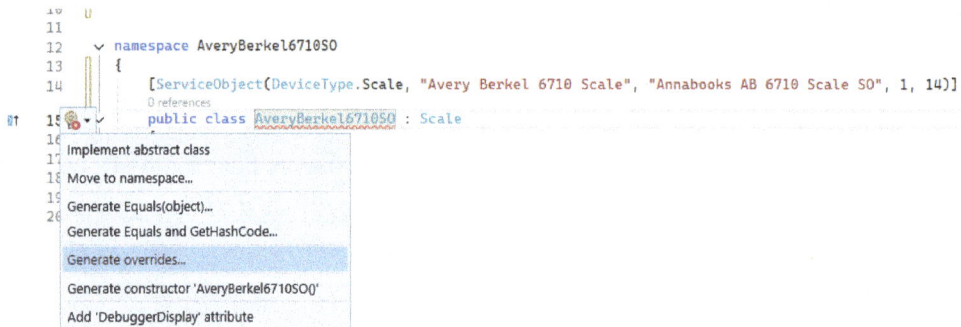

4. Select "Implement abstract class. All the properties and methods from the ScaleBasic class are added to the library.

114

5. Now we are ready to add our code. Just within the public class AveryBerkel6710SO : Scale, add the following to define the serial port with the default COM1 port. This will become generic later when we fill in the Open() method.

```
public class AveryBerkel6710SO : Scale
{

    protected SerialPort sPort = new SerialPort("COM1", 9600,
Parity.Even, 7, StopBits.One);

    public         override       int      DataCount      =>      throw      new
NotImplementedException();
```

6. Since there so many properties and methods to fill in, we will fill in the code by group. The first step is to fill in the capabilities (Cap). The Scale Service Object requires various capabilities / properties to be completed. The user manual for the scale lists the features available. The 6710 is a very simple scale so most scale capabilities are not available, but return calls must be made. There is no tare weight, price calculations, text output display, device events, data events, etc. The display only displays the weight, and it cannot be programmed. The scale does allow the program to zero the scale, so this capability is set to true. The following code shows all the properties:

```
public override bool AutoDisable
{
    get
    {
```

```
            return false;
        }
    set
    {
        throw new NotImplementedException();
    }
}
public override bool CapDisplay { get { return false; } }

public override bool CapDisplayText { get { return false; } }

public override bool CapPriceCalculating { get { return false; }
}

public override bool CapTareWeight { get { return false; } }

public override bool CapZeroScale { get { return true; } }

public override bool AsyncMode
{
    get
    {
        return false;
    }
    set
    {
        throw new NotImplementedException();
    }
}

public override int MaxDisplayTextChars
{
    get { return 0; }
}

public override decimal MaximumWeight
{
    get { return 30; }
}

public override decimal SalesPrice
{
```

```csharp
        get { return 0; }
}

public override decimal TareWeight
{
    get
    {
        return 0;
    }
    set
    {
        throw new NotImplementedException();
    }
}
public override decimal UnitPrice
{
    get
    {
        return 0;
    }
    set
    {
        throw new NotImplementedException();
    }
}

public override WeightUnit WeightUnit
{
    get { return WeightUnit.Pound; }
}

public override bool ZeroValid
{
    get
    {
return false;
    }
    set
    {
        throw new NotImplementedException();
    }
}
```

```csharp
public override PowerReporting CapPowerReporting
{
    get { throw new NotImplementedException(); }
}

public override bool CapStatisticsReporting
{
    get { return false; }
}

public override bool CapUpdateStatistics
{
        get { return false; }
}
```

7. The CheckHealth and CheckHealthText must be filled out. The CheckHealth method performs the status 'S' command / Get Status on the scale. There are different types of Health checks: External, Interactive, and Internal. For this scale, there is only Internal, but the code won't check for the level. When the code behind the Scale Status button is added, all the error conditions will be sent back as strings. The test application returned user-friendly strings, which is nice for testing. The UnifiedPOS specification calls out the general error return codes, and POS for .NET already has these general codes defined. Using the SDK help file, you have to line up the specific error with the error from the scale. For example, a scale in motion should return the StatusWeightUnstable as a string. The code below lists some of the error responses. The manual lists more possibilities. The CheckHealthText could be something more, but a simple call to the CheckHealth method is made.

```csharp
public override string CheckHealth(HealthCheckLevel level)
{
    Char[] theStatus = new char[7];
    String FinalStatus = "ok";

    sPort.Write("S" + (char)13);
    Thread.Sleep(1000);

    try
    {
        theStatus                                              =
sPort.ReadTo(((char)3).ToString()).ToArray();
    }
    catch
```

118

```csharp
    {
        FinalStatus = StatusNotReady.ToString();
    }


    if ((theStatus[1].ToString() + theStatus[2].ToString() +
theStatus[3].ToString()) == "S10")
    {
        FinalStatus = StatusWeightUnstable.ToString(); //scale
is in motion
    }
    else if ((theStatus[1].ToString() + theStatus[2].ToString()
+ theStatus[3].ToString()) == "S20")
    {
        FinalStatus = StatusWeightZero.ToString();   //scale is
at zero and stable
    }
    else if ((theStatus[1].ToString() + theStatus[2].ToString()
+ theStatus[3].ToString()) == "S00")
    {
        FinalStatus = StatusStableWeight.ToString(); //Scale Not
at Zero and Stable
    }
    else if ((theStatus[1].ToString() + theStatus[2].ToString()
+ theStatus[3].ToString()) == "S02")
    {
        FinalStatus =        ExtendedErrorOverWeight.ToString();
//Scale over Max Weight
    }


    return FinalStatus;
}


public override string CheckHealthText
{
    get
    {
        return CheckHealth(HealthCheckLevel.Internal);

    }
}
```

8. The ReadWeight method gets the weight result from the scale. There are some changes from the test application to address the UnifiedPOS specification. The last characters from the original weight code include the abbreviations for the unit of measure. Since only a decimal value is to be returned, these last two characters for the unit of measure are dropped from the final weight value. If there is an error, -1 is returned so that the application can flag and handle the error. A flag is needed, since an error code could be mistaken for weight value.

```
public override WeightTareInfo ReadLiveWeightWithTare(int weightData,
int tare, int timeout)
{
    throw new NotImplementedException();
}

public override decimal ReadWeight(int timeout)
{
    Char[] theWeight = new char[16];
    Decimal FinalWeight = 0;
    sPort.Write("W" + (char)13);
    Thread.Sleep(1000);

    try
    {
        theWeight = sPort.ReadTo(((char)3).ToString()).ToArray(); //Read to
the end of the line ETX
    }
    catch
    {
        return -1;
    }

    if     ((theWeight[1].ToString()     +     theWeight[2].ToString()     +
theWeight[3].ToString()) == "S10")
    {
        return -1;
    }
    else   if   ((theWeight[1].ToString()     +   theWeight[2].ToString()   +
theWeight[3].ToString()) == "S11")
    {
        return -1;
    }
    else
    {
```

120

FinalWeight = Decimal.Parse(theWeight[1].ToString() + theWeight[2].ToString() + theWeight[3].ToString() + theWeight[4].ToString() + theWeight[5].ToString() + theWeight[6].ToString());
return FinalWeight;
}
}

9. The ZeroScale method doesn't return anything. The status checks in the test code don't apply. Here is the code for this method:

```
public override void ZeroScale()
{
        sPort.Write("Z" + (char)13);
        Thread.Sleep(1000);
}
```

10. Add some descriptive details in the Service Object for when POSDM or SOManager calls for details.

```
public override string ServiceObjectDescription
{
    get { return "Scale"; }
}
```

```
public override string DeviceName
{
    get { return "Avery Berkel 6710"; }
}
```

```
public override string ServiceObjectDescription
{
    get { return "Scale"; }
}
```

11. Finally, we need to add Overrides for the Open, Close, Claim, and Release methods for the opening and closing of the COM port.

```
public override void Open()
{
    sPort.PortName = DevicePath;
    sPort.Open();
}
```

```
public override void Close()
{
    sPort.Close();
}

public override bool Claimed
{
    get { return true; }
}

public override void Claim(int timeout)
{
    //Visual Studio will not build if the method is missing.
    int x = 0;
    timeout = x + 100;
}

public override void Release()
{
    //Visual Studio will not build if the method is missing.
    int x = 0;
    x++;
}
```

For the Open method, DevicePath is part of the PosCommon class. The path must be set using POSDM or SOManager to a COM port or COM1 will be used. The Close method, simply closes the port. The Claimed property method simply reports true, while the Claim method has some dummy code because the libdrart will not build without the code. The Release method also has some dummy code.

12. Save the project.
13. Open AssemblyInfo.cs to modify the assembly information as follows. First, add the reference if they are not already added:

```
using System.Reflection;
using System.Runtime.CompilerServices;
using System.Runtime.InteropServices;
using System;
using System.Security.Permissions;
using Microsoft.PointOfService;
using Microsoft.PointOfService.BasicServiceObjects;
```

14. Next, in the General information assembly section change the information to match your company.

```
[assembly: AssemblyTitle("AveryBerkel6710SO")]
[assembly: AssemblyDescription("")]
[assembly: AssemblyConfiguration("")]
[assembly: AssemblyCompany("Annabooks")]
[assembly: AssemblyProduct("AveryBerkel6710SO")]
[assembly: AssemblyCopyright("Copyright ©  2024 Annabooks LLC")]
[assembly: AssemblyTrademark("")]
[assembly: AssemblyCulture("")]
```

15. Change the assembly version information to match UnifiedPOS version

```
[assembly: AssemblyVersion("1.14.1.0")]
[assembly: AssemblyFileVersion("1.14.1.0")]
```

16. At the end of the file, we need to add some more details to access the assembly.

```
[assembly: ClassInterface(ClassInterfaceType.None)]
[assembly: CLSCompliant(true)]
```

```
[assembly: AssemblyDelaySign(false)]
[assembly: AssemblyKeyFile("")]
[assembly: AssemblyKeyName("")]
[assembly: PosAssembly("Annabooks")]
```

17. Save the project.
18. Build the Service Object. Correct any errors. Typically, reminders will appear for some things to be filled out that don't need filling out.

5.4.3   Part 3: Set Up the Service Object

We will set the registry path to point to the debug version of the Service Object. Service Object Manager will be used to check the installation and set up a logical name. The TestApp.exe will be used to perform a basic test.

1. Open the Regedit utility in Administrator mode.
2. Under HKLM\SOFTWARE\Wow6432Node\POSfor.NET\ControlAssemblies, add a new string value called ABSCALE.
3. Set the Value data to point to the path of the Service Object dll. For test and development, it is a good practice to point to the debug version.

OW6432Node\POSfor.NET\ControlAssemblies

Name	Type	Data
ab (Default)	REG_SZ	C:\Program Files (x86)\Common Files\Microsoft Shared\Point Of Service\Control Ass...
ab EPSONSOs	REG_SZ	C:\NETPOS\\OPOS for .NET
ab ExampleSOs	REG_SZ	C:\Program Files (x86)\Microsoft Point Of Service\SDK\Samples\Example Service Obje...
ab Simulators	REG_SZ	C:\Program Files (x86)\Microsoft Point Of Service\SDK\Samples\Simulator Service Obj...
ab ABSCALE	REG_SZ	C:\NETPOS\AveryBerkel6710SO\AveryBerkel6710SO\bin\Debug\AveryBerkel6710SO.dll

4. Now, we can use Service Object Manager to see if the Service Object is working. Open Service Object Manager in Administrator mode.
5. Click on the Get POS Devices button:

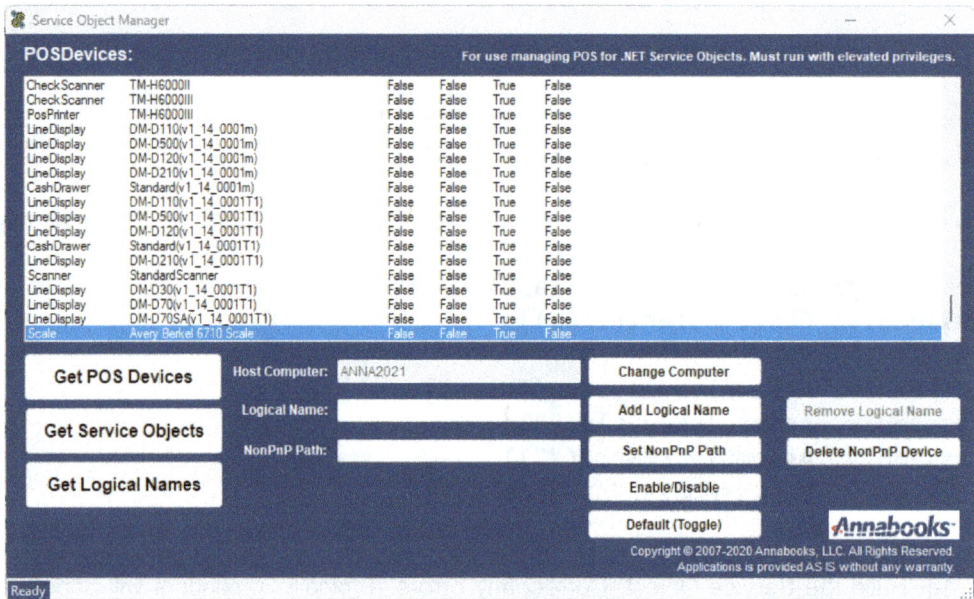

6. A logical name can be created for the scale. Click on the Scale in the list box.
7. Enter ABScale in the Logical Name text box, and click on the Add Logical Name button.

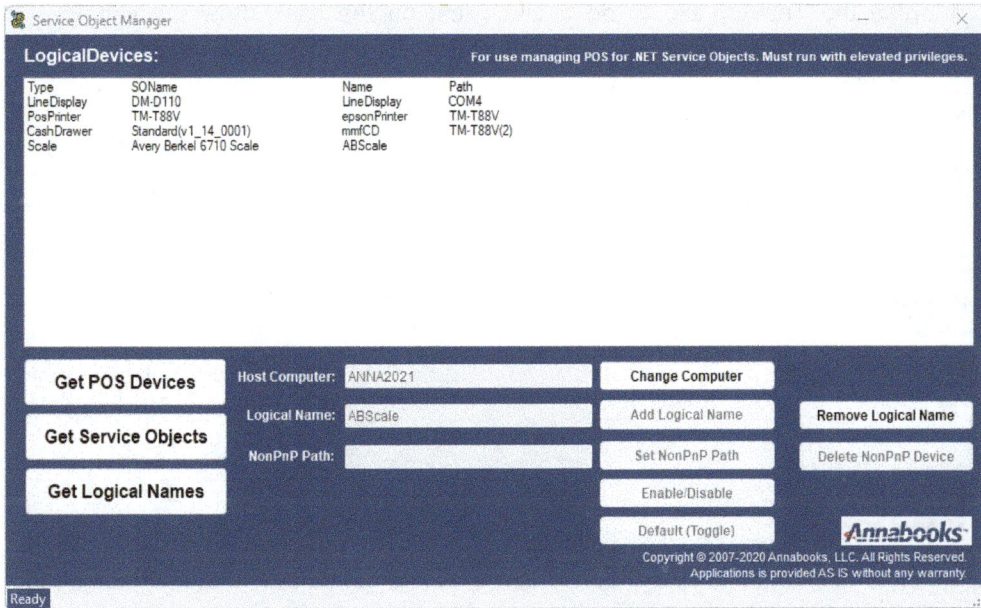

Service Object Manager

LogicalDevices: For use managing POS for .NET Service Objects. Must run with elevated privileges.

Type	SOName		Name	Path
LineDisplay	DM-D110		LineDisplay	COM4
PosPrinter	TM-T88V		epsonPrinter	TM-T88V
CashDrawer	Standard(v1_14_0001)		mmfCD	TM-T88V(2)
Scale	Avery Berkel 6710 Scale		ABScale	

Host Computer: ANNA2021   Change Computer

Get POS Devices

Logical Name: ABScale   Add Logical Name   Remove Logical Name

Get Service Objects

NonPnP Path: Set NonPnP Path   Delete NonPnP Device

Get Logical Names   Enable/Disable

Default (Toggle)   *Annabooks*

Copyright © 2007-2020 Annabooks, LLC. All Rights Reserved.
Applications is provided AS IS without any warranty.

Ready

8. The final step is to set the COM port Path for the scale. Click on Get POS Devices.
9. Click on the Avery Berkel Scale services object.
10. Enter COM# for the COM port that the device is connected to in the NonPnP path text box and click on the Set NonPnP Path button.

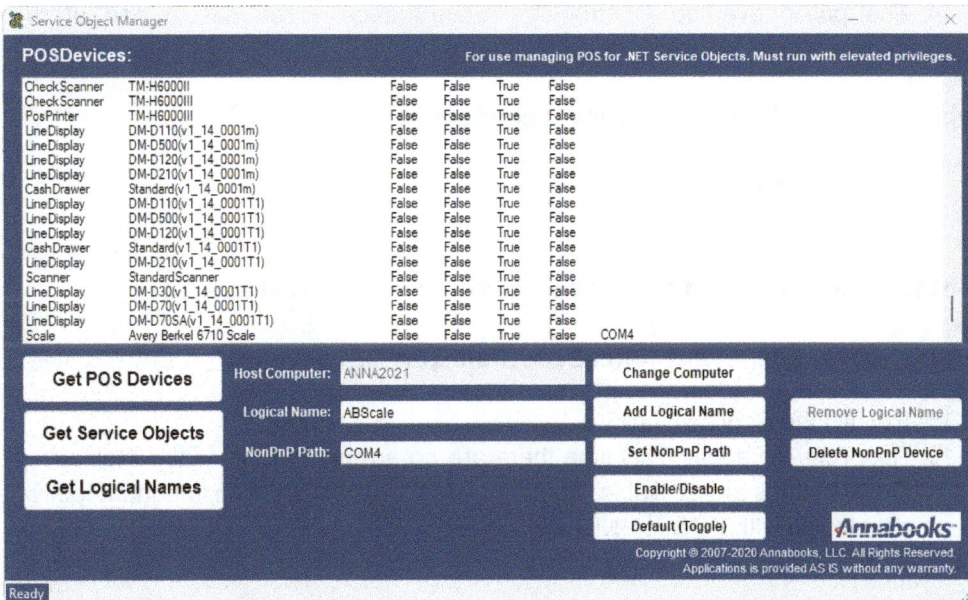

Service Object Manager

POSDevices: For use managing POS for .NET Service Objects. Must run with elevated privileges.

CheckScanner	TM-H6000II	False	False	True	False	
CheckScanner	TM-H6000III	False	False	True	False	
PosPrinter	TM-H6000III	False	False	True	False	
LineDisplay	DM-D110(v1_14_0001m)	False	False	True	False	
LineDisplay	DM-D500(v1_14_0001m)	False	False	True	False	
LineDisplay	DM-D120(v1_14_0001m)	False	False	True	False	
LineDisplay	DM-D210(v1_14_0001m)	False	False	True	False	
CashDrawer	Standard(v1_14_0001m)	False	False	True	False	
LineDisplay	DM-D110(v1_14_0001T1)	False	False	True	False	
LineDisplay	DM-D500(v1_14_0001T1)	False	False	True	False	
LineDisplay	DM-D120(v1_14_0001T1)	False	False	True	False	
CashDrawer	Standard(v1_14_0001T1)	False	False	True	False	
LineDisplay	DM-D210(v1_14_0001T1)	False	False	True	False	
Scanner	StandardScanner	False	False	True	False	
LineDisplay	DM-D30(v1_14_0001T1)	False	False	True	False	
LineDisplay	DM-D70(v1_14_0001T1)	False	False	True	False	
LineDisplay	DM-D70SA(v1_14_0001T1)	False	False	True	False	
Scale	Avery Berkel 6710 Scale	False	False	True	False	COM4

Host Computer: ANNA2021   Change Computer

Get POS Devices

Logical Name: ABScale   Add Logical Name   Remove Logical Name

Get Service Objects

NonPnP Path: COM4   Set NonPnP Path   Delete NonPnP Device

Get Logical Names   Enable/Disable

Default (Toggle)   *Annabooks*

Copyright © 2007-2020 Annabooks, LLC. All Rights Reserved.
Applications is provided AS IS without any warranty.

Ready

11. Use the TestApp.exe to do a quick test on the cash draw functionality. The TestApp.exe is always a good application to test the installation of a Service Object.

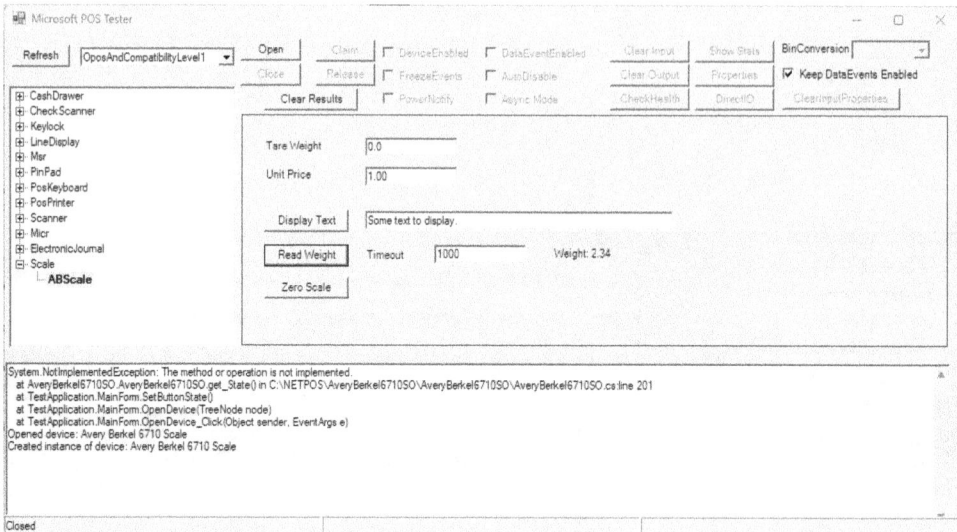

12. The first exception is complaining about a method not being implemented, but the system still opens the device. Close the TestApp.exe.
13. Back to Visual Studio and the AveryBerkel7710.cs file, go to line 201.
14. The public override ControlState State needs to be set to one of the 4 enumerations. Let's set it to idle.

```
public override string ServiceObjectDescription
{
    get { return "Scale"; }
}

public override ControlState State => ControlState.Idle;

public override event DataEventHandler DataEvent;
```

15. Rebuild the Service Object.
16. Run TestApp again. This time there are no errors, and the Claim, Release, and Close buttons are available. Again, the Test application is an important tool when developing with POS for .NET.

126

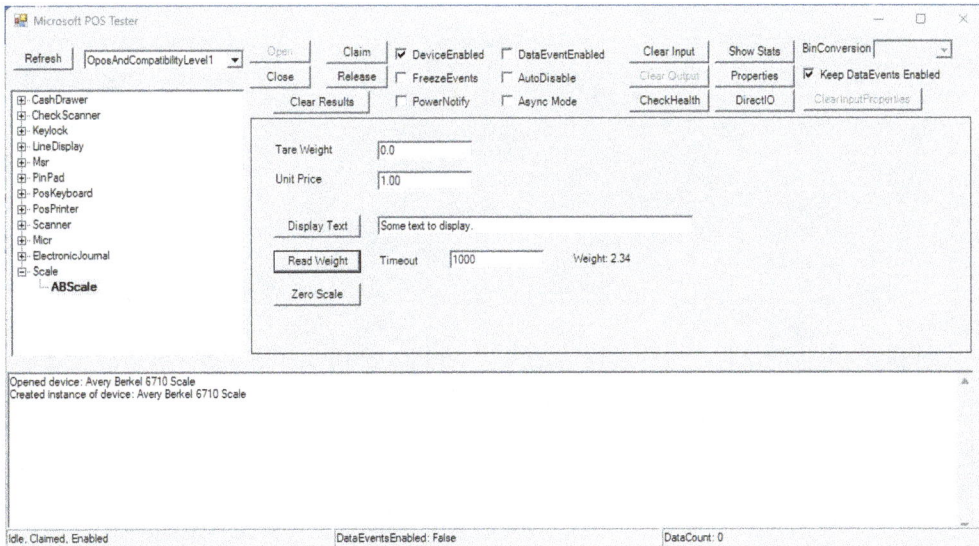

5.5   *Exercise 5.3 Scale Application*

With the new scale Service Object almost ready, we can create an application to test against it and even debug into the Service Object.

5.5.1   Part 1: Create the Application and Form

The first step is to create the project. The application will look similar to the test application from earlier in the chapter.

1. Open Visual Studio.
2. Click on "Create a new Project".
3. Set the language to C# and search for the "Windows Forms App (.NET Framework)" template:

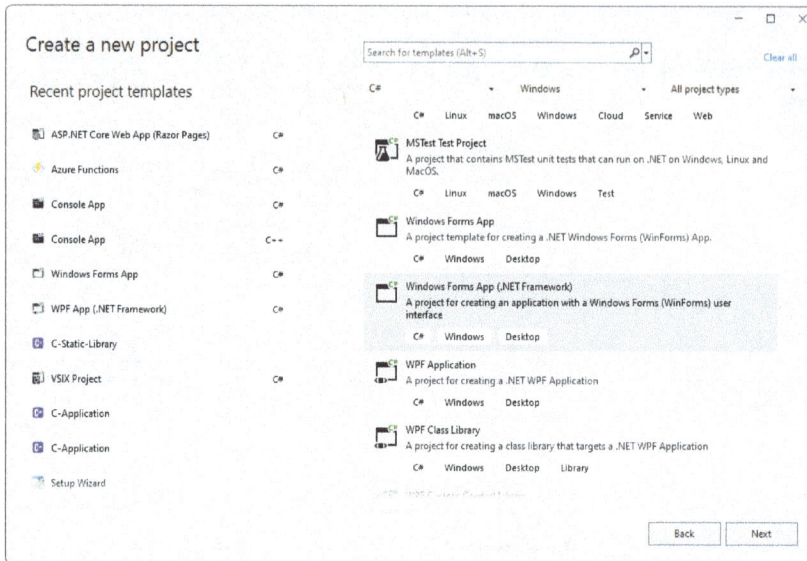

4. Click Next.
5. Name the project EX53_Scale_App.
6. Set the location for the project.
7. Keep the Framework set to .NET Framework 4.8.
8. Click Create.
9. On the form, add the following controls and the properties:

> Label:
> > Name: lblWeight
> > Text: Weight:
> > Font: Arial, Regular, 24pt

> Label:
> > Name: lblWeightResult
> > Text: 0.00
> > Font: Arial, Regular, 24pt

> Button:
> > Name: btnGetWeight
> > Text: Get Weight
> > Font: Arial, Regular, 12pt

> Button:
> > Name: btnZeroScale
> > Text: Zero Scale
> > Font: Arial, Regular, 12pt

Button:
  Name: btnGetStatus
  Text: Scale Status
  Font: Arial, Regular, 12pt

StatusStrip:
  Click the drop-down to add: StatusLabel
    Name: TTStatus
    Text: Ready
    Font: Arial, Regular, 12pt

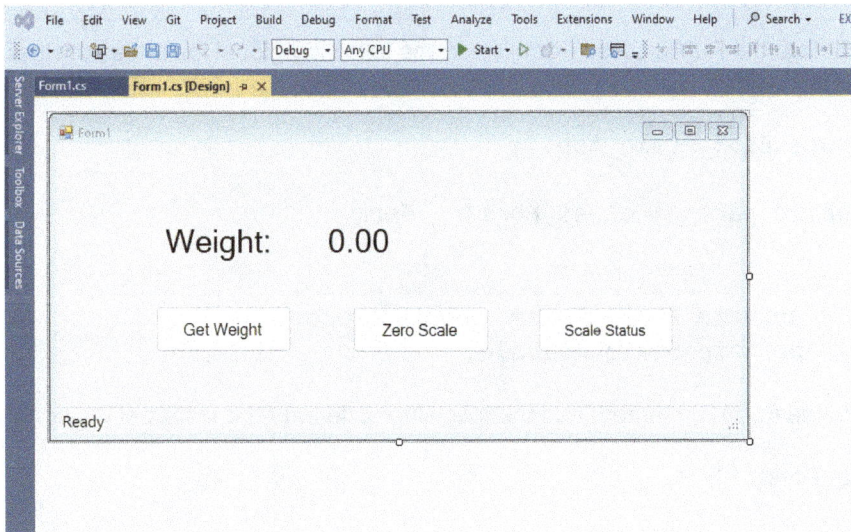

10. Save the project.

5.5.2   Part 2: Adding the Code
With the UI created, let's fill in the code.

1. The first step is to add the Microsoft.PointOfService.dll reference. From the menu, select Project->Add Reference. This will open the Add Reference dialog.
2. If you previously added the reference in the last exercise, all you have to do is check the box next to Microsoft.PointOfService.dll and click OK. Otherwise, click on the Browse tab, and locate the Microsoft.PointOfService.dll found under "C:\Program Files\Microsoft point of Service\SDK".
3. Click add.
4. Click on the OK button.
5. Open Form1.CS in code view.
6. At the top of the code, before the From1 class, add the Microsoft.PointOfService imports:

```
using System;
using System.Collections.Generic;
using System.ComponentModel;
using System.Data;
using System.Drawing;
using System.Linq;
using System.Text;
using System.Threading.Tasks;
using System.Windows.Forms;
using Microsoft.PointOfService;
```

7. Before the Public Class Form1, add the following global instances of PosExplorer and Scale.

```
namespace EX53_Scale_App
{
    public partial class Form1 : Form
    {

        private PosExplorer myPosExplorer;
        private Scale myScale;
```

8. In the Form1() method, add the following code after the InitializeComponent() call.

```
public Form1()
{
    InitializeComponent();
    myPosExplorer = new PosExplorer();
    DeviceInfo    device    =    myPosExplorer.GetDevice("Scale",
"ABScale");

    if (device == null) {

        TTStatus.Text = "Scale not found";


    }
    else
    {
        myScale = (Scale)myPosExplorer.CreateInstance(device);
        myScale.Open();
        myScale.Claim(1000);
        myScale.DeviceEnabled = true;
        TTStatus.Text = "Scale Ready!";
```

```
        }
    }
```

Technically, Claim() and setting the DeviceEnable property are not required as the Service Object just hardcoded these values as true. The method and property in the Service Object could do more, but the device is simple and doesn't need anything more sophisticated. Since this is a serial device, there is no DeviceAdded or DeviceREmvoed events, nor is there a DataEvent triggered by the scale.

9. Go back to Designer mode, and double-click on each button to generate the button handlers for each button.
10. Add the code for the btnGetWeight_Click handler method:

```
private void btnGetWeight_Click(object sender, EventArgs e)
{
    Decimal theWeight = 0;
    String ScaleStatus;

    try
    {
        theWeight = myScale.ReadWeight(1000);
    }
    catch
    {
        theWeight = 0;
        TTStatus.Text = "Error";
    }

    if (theWeight.ToString() == "-1")
    {

        ScaleStatus                                        =
myScale.CheckHealth(HealthCheckLevel.Internal);
        switch (ScaleStatus)
        {
            case "11":
                TTStatus.Text = "Scale Stable and not at Zero";
                break;

            case "12":
                TTStatus.Text = "Scale is in motion and not
ready";
                break;
```

```
        case "13":
            TTStatus.Text = "Scale is at Zero and Stable";
            break;

        default:
            TTStatus.Text = "Unkown Error";
            break;
        }
    }
    else
    {
        lblWeightResult.Text = theWeight.ToString() + " lbs";
        TTStatus.Text = "Ready";
    }
}
```

The Service Object is going to return a weight value in decimal or return a -1. If a weight value is returned, the value will be printed to the screen. If -1 is returned, the CheckHealth status is called to determine the issue.

11. Add the code for the btnZeroScale_Click handler method:

```
private void btnZeroScale_Click(object sender, EventArgs e)
{
    myScale.ZeroScale();
    TTStatus.Text = "Zero Command called, check scale display";
}
```

12. Add the code for the btnGetStatus_Click handler method:

```
private void btnGetStatus_Click(object sender, EventArgs e)
{
    String                      ScaleStatus                     =
myScale.CheckHealth(HealthCheckLevel.Internal);
    switch (ScaleStatus)
    {
        case "11":
            TTStatus.Text = "Scale Stable and not at Zero";
            break;

        case "12":
            TTStatus.Text = "Scale is in motion and not ready";
```

```
        break;

    case "13":
        TTStatus.Text = "Scale is at Zero and Stable";
        break;

    default:
        TTStatus.Text = "Unkown Error";
        break;
    }
}
```

13, Save the project.

5.5.3   Part 3: Build and Test
With the code added, let's test the application.

1. Make sure the scale is connected to the computer via the serial port or USB-to-serial adapter.
2. Double-check that the COM port in Device Manager matches.
3. Change the project properties and uncheck "Prefer 32-bit" from the Build options.

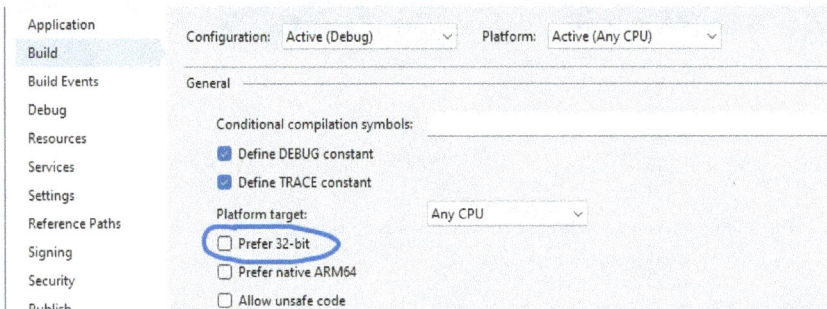

4. Start the application by hitting F5 or selecting Debug->Start Debugging from the menu.
5. Click on the Get Weight button, and the weight should be returned.

Scale Stable and not at Zero

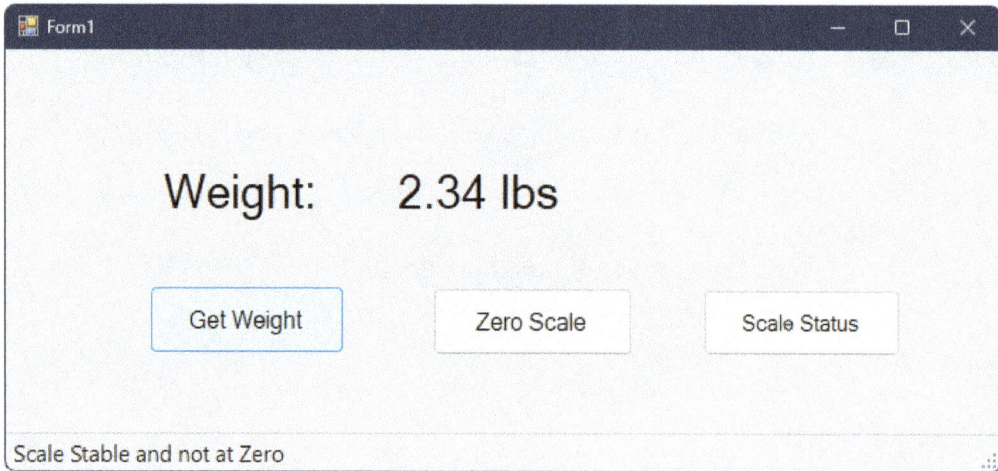

6. In Visual Studio, set a breakpoint on the line with "theWeight = myScale.ReadWeight(1000);"
7. Click on the "Get Weight" button.
8. The code will stop at the breakpoint. Click on the Step-Into button, and Visual Studio will jump into the Service Object source code where you can debug in the Service Object.

9. Click Continue to let the application run.
10. Stop debugging when finished.

The final step would be to build the Scale Service Object as a release build. For deployment, you can put the Service Object in the "C:\Program Files (x86)\Common Files\Microsoft Shared\Point Of Service\Control Assemblies" or any other location so long as the registry key path

(HKLM\SOFTWARE\Wow6432Node\POSfor.NET\ControlAssemblies) has an entry point to the DLL location.

The Service Object has several advantages:

- Offloads accessing the hardware from the application to a library. Thereis no need to hardcode COM ports or manage a serial port instance.
- Take advantage of the POSDM and the WMI API to configure the Service Object for a logical name and a COM port path.
- The application can use the standard POS for .NET methods to access the Service Object and interact with the device.

5.6 Exercise 5.3 Modify the Example MSR SO for Hardcode ID

The scale Service Object is a non-Plug and Play device, and a COM# was required to be set for the Device Path. Chapter 2 demonstrated how to set up a Plug and Play device Service Object using an XML file. You may notice some Service Objects don't use XML files. That is because the Hardware ID is hardcoded into the Service Object. In this exercise, we will modify the Example MSR from the POS for .NET SDK to hardcode the Hardware ID into the Service Object.

5.6.1 Part 1: Get the Hardware IDs

Again, here are the steps to get the Hardware ID for the USB device.

1. If plugged in, unplug the MSR USB device from the system.
2. Open Control Panel.
3. Open the System Control Panel applet.
4. Click on the Hardware tab.
5. Click on the Device Manager button.
6. Plug the MSR USB device into an open USB port. Device Manager should update with two new devices listed under Human Interface Devices.
7. Open the Properties for the HID-compliant Devices for the MSR.
8. Click on the Details tab.
9. In the drop-down, select Hardware Ids.

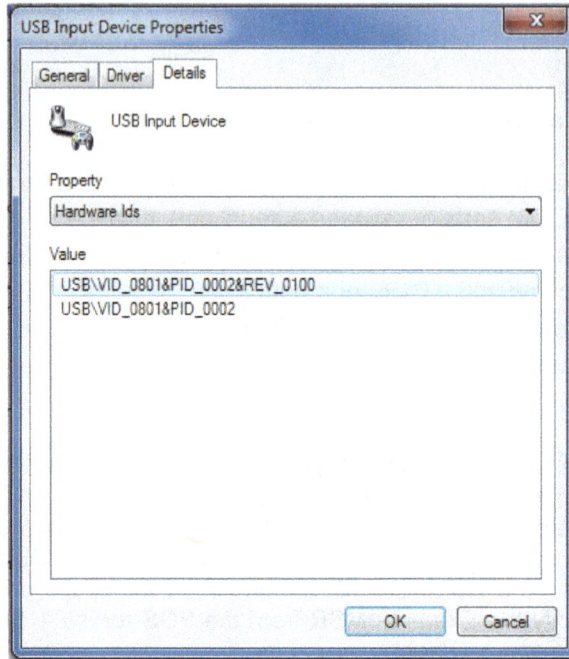

The hardware IDs for this example are:

- HID\Vid_0801&Pid_0002
- HID\Vid_0801&Pid_0002&Rev_0100

5.6.2   Part 2: Adding the HardwareIDAttribute

Let's rebuild the Example Service Object with the MSR PnP information:

1. Copy the Samples Folder SDK project found at: "C:\Program Files(x86)\Microsoft Point Of Service\SDK" to a different location on your system (i.e. C:\NETPOS).
2. Open Visual Studio.
3. Open the Example Service Object project. You may be asked to convert the project to Visual Studio. Choose the defaults until finished.
4. Open the ExampleMsr.CS file.
5. Expand the ExampleMsr Class region.
6. Between the ServiceObject and the Public class ExampleMsr : MsrBase, enter the following:

[HardwareId("HID\\Vid_0801&Pid_0002", "HID\\Vid_0801&Pid_0002")]

```
25
26   v   namespace Microsoft.PointOfService.ExampleServiceObjects
27   |   {
28
29   v       #region ExampleMsr Class
30   |       [ServiceObject( DeviceType.Msr,
31   |                       "ExampleMsr",
32   |                       "Service object for Example MSR", 1, 14)]
33   |       [HardwareId("HID\\Vid_0801 & Pid_0002", "HID\\Vid_0801 & Pid_0002")]
             2 references
34   |v      public class ExampleMsr : MsrBase
35   |       {
```

7. Save the project.
8. Build the project by clicking Build->Build Solution. You may get an error accessing the registry to set the location of the DLL. This can be ignored.
9. Remove any XML configuration file from the "C:\Program Files (x86)\Common Files\microsoft shared\Point Of Service\Control Configurations" directory.
10. Now, we are going to swap the original Service Object with the one that we just built. Copy the Microsoft.PointOfService.ExampleServiceObjects.dll found under "C:\Program Files(x86)\Microsoft Point Of Service\SDK\Samples\Example Service Objects" to a new folder called "C:\Program Files\Microsoft Point Of Service\SDK\Samples\Example Service Objects\Backup".
11. Copy the recently built Microsoft.PointOfService.ExampleServiceObjects.dll from "C:\Program Files(x86)\Microsoft Point Of Service\SDK\Samples\Example Service Objects\Source\bin\Debug" to "C:\Program Files(x86)\Microsoft Point Of Service\SDK\Samples\Example Service Objects".
12. Plug in the MSR to see that the application detects the new device.
13. Run the MSR application created in Chapter 2, or run the TestApp.
14. Swipe a card through the MSR to verify that the data gets read.
15. Replace the original Microsoft.PointOfService.ExampleServiceObjects.dll to the "C:\Program Files (x86)\Microsoft Point Of Service\SDK\Samples\Example Service Objects" folder.

5.7 Three Methods to Setup a Service Object

We now have three methods to setup a Service Object as shown in the table below:

Device Type	Setup Option
PnP	XML File in "C:\Program Files(x86)\Common Files\Microsoft Shared\Point Of Service\Control Configurations"
PnP	Hardcoded ID within the Service Object [HardwareId("From", "To")]
Non-PnP	POSDM or WMI to set the DevicePath and logical name

The final folder location is up to the developer. Either place the Service Object DLL in the POS for .NET predefined folder location (C:\Program Files (x86)\Common Files\Microsoft Shared\Point Of Service\Control Assemblies), or place the assembly in a custom path and set the registry keys to point to that path.

5.8   What about Pin Pads?

One of the big questions that appears on forums is POS for .NET and Pin Pads. Most checkout locations have a multifunctional pin pad that has the following: NRF tap, smart card reader, MSR, and signature pad. These devices require special software provided by the OEM to customize, which will include a custom API for access. Any OPOS driver or Service Object that might be available would have to come from the POS device manufacturer.

5.9 Summary: Service Objects – The Key to Write Once, Support Many

Abstracting the hardware from the software is the cornerstone of the UnifiedPOS specification and the POS for .NET implementation. Application developers can write one application with POS for .NET and have it support multiple POS devices. Like OPOS drivers, Service Objects are the key to making "write once, support many" a reality. POS for .NET greatly reduces the complexity of developing Service Objects, as was demonstrated for the scale device. Care must be taken when using Service Objects and OPOS drivers in the same system. Devices sharing the same physical port resource need to be addressed programmatically in the application. This chapter concludes the POS for .NET discussion for now. The next two chapters cover .NET Core applications using the Universal Windows POS assemblies, and we will see why POS for .NET is still important for use with .NET Core applications.

6   .NET and POS WinRT

As Azure exploded on the scene over a decade ago, Microsoft needed to change the thought process on Windows. Rather than just supporting the Windows operating system, Microsoft extended Visual Studio to support different operating systems like iOS and Linux so that all these platforms could connect to Azure. The programming paradigm also shifted. As Microsoft made one last push into the mobile space, they developed a new programming paradigm called Universal Windows Programs (UWP) built-in Windows Runtime (WinRT), where UWP applications could be downloaded from the Microsoft store. The history and fate of Windows Mobile and UWP are well known. Windows Mobile failed. Windows 8 and the new shell interface were not welcome for desktop PCs. UWP looked good on paper, but it was not a practical programming solution. Clearly, a learning process and one that business schools will cover in detail. Microsoft was able to pivot and develop .NET core or simply .NET. The .NET runtime supports Intel and ARM processors, as well as, a variety of operating systems: Windows, iOS, Linux, etc. .NET Framework will continue to be part of Windows, but it will not be advanced as much as .NET will be moving forward.

In the journey from UWP to .NET, a new way for UWP applications to access hardware was created. To allow POS application developers to create applications with UWP and WinRT, Microsoft created Universal Windows (UW) POS assemblies to support POS hardware using the new enumeration methods. The new mechanism to access hardware is very different from the createinstance() method used by POS for .NET. This chapter will focus on the following:

- The new POS device connection and action sequence
- The available POS drivers that are supported in Windows
- Writing .NET applications to access the POS WinRT namespace using the different enumeration methods

6.1 Device Enumeration Methods

Per the previous chapters, we have seen that a POS for .NET application would use POS Explorer and DeviceInfo to get the handle to the device, which could then be used to create an instance. Once the instance is created, you can perform the Open, Claim, Action, Release, and Close operations on the device. Service Objects and OPOS drivers are not Windows device drivers. POS for .NET acted as a device manager so POS applications could be developed following the UnifiedPOS specification.

The UW POS assemblies are actually Windows device drivers. Windows device management plays a critical role in managing these devices. For a device connected locally via USB or serial, the drivers will show up in Device Manager, thus there is no need for the POSDM utility or WMI API to set logical names or paths. For devices connected remotely via Ethernet or Bluetooth, a device gets paired with the Windows operating system. The operation to connect to such a POS device is different from POS for .NET. Here is the new sequence:

- A POS device query, which can search for devices locally or remotely, is used to locate devices of a specific device class. The query uses an Advanced Query Syntax (AQS) device interface selector string that returns information about an instance of the DeviceInformation() class.
- The Device ID property that is in the DeviceInformation instance is used to find a specific device using the FindAllAsync() method.
- Once the specific device has been found, a claim can be performed.
- If the claim is successful, the application can perform actions on the POS device.
- When finished a Dispose method is called to release the claimed device.

POS for .NET   WinRT

DeviceInfo – GetDevice from POS Explorer
CreateInstance
Open
Claim
POS Device Action
Release
Close

Query for Device
Get Device Information (device ID)
Connect to the POS device using device ID
Claim
POS Device Action
Dispose

4 Enumeration Methods

The query can come back with more than one device of the specified device class. For example, a query for a barcode scanner in the TabletKiosk TufTab i60XT or i61J will return the built-in scanner, as well as, the front and back cameras. The POS application needs to be able to address such cases. For UWP and WinRT, four enumeration methods were developed to establish the connection to the device.

1. Device Picker Method – Provides a flyout with a list of available devices for a specific device class from which the user can select. The list gets dynamically updated as devices are connected and removed from the system. The picker method is limited to UWP applications.
2. Get the first available device using GetDefaultAsync() method – When the method is called, the first device ID for the queried device class is returned. If there is more than one device of a device class that is ready, the first device returned may not be the desired device. This enumeration method should only be used when you know that there is only one device of a device class available. The method will not work for remotely connected devices, since device pairing has to take place.
3. Snapshot of devices – This enumeration method gets a snapshot of the available devices at the time of the call. Developers can create their own device UI or enumerate devices without displaying a UI.

4. Device Watcher – This enumeration method is close to what is done with POS for .NET. It enumerates devices that are currently present and then monitors for devices that are added or removed in the background.

6.2 POS WinRT Namespaces and Universal Windows Assembly Support

POS for .NET was provided as a separate SDK that had a runtime engine, tools, and an API library to create applications and manage POS devices. POS for .NET matched the support found in UnifiedPOS including support for 36 POS device types.

Support for UWP POS applications was introduced in Windows 10 Build 1607 version 14393 as part of the WinRT. With WinRT, a separate monitoring runtime or SDK is not needed. The Windows SDK for that release added the following namespaces to create UWP applications:

- Windows.Devices.PointOfService – the main namespace that contains the classes for accessing and interacting with POS devices.
- Windows.Devices.Enumeration – Classes for connecting devices to applications.
- Windows.Security.Cryptography - Classes for supporting several 1D, 2D, and QR barcode symbologies.

That is all good and fine for UWP, but what about .NET Core? For .NET Core 3.1, a NuGet package (Microsoft.Windows.SDK.Contracts.) was needed to access the UWP /WinRT POS namespaces. As .NET has grown, much of the support has been made available to .NET without the need for a separate NuGet package, but not everything has been brought over.

As POS for .NET, Service Objects, and OPOS drivers act like the abstraction between hardware and application. The UW assemblies are the equivalent abstraction for WinRT. For WinRT, Microsoft created full-production UW assemblies for the five popular POS device types:

POS Device	Connection Type	Driver
Barcode Scanner	USB, Bluetooth, Webcam	Built-in UMDF driver
Cash Drawer	Network/Bluetooth, DK port, and OPOS	Built-in UMDF driver and OPOS
Line Display	OPOS	OPOS
Magnetic Stripe Reader (MSR)	USB	Built-in UMDF driver
Receipt Printer	Network/Bluetooth and OPOS	Built-in UMDF driver and OPOS

If these look familiar, these five devices make up the five important classes in the Windows.Devices.PointOfService namespace: BarcodeScanner, CashDrawer,

142

LineDisplay, MagneticStripeReader, and POSPrinter. In total, there are 67 classes, 6 interfaces, 1 structure, and 37 enums that make up the Windows.Devices.PointOfService namespace. For the remaining 31 devices in the UnifiedPOS specification, these device types can be accessed using desktop bridge technology that allows .NET and .NET Framework applications to communicate with one another, thus everything that you learned with POS for .NET is not lost.

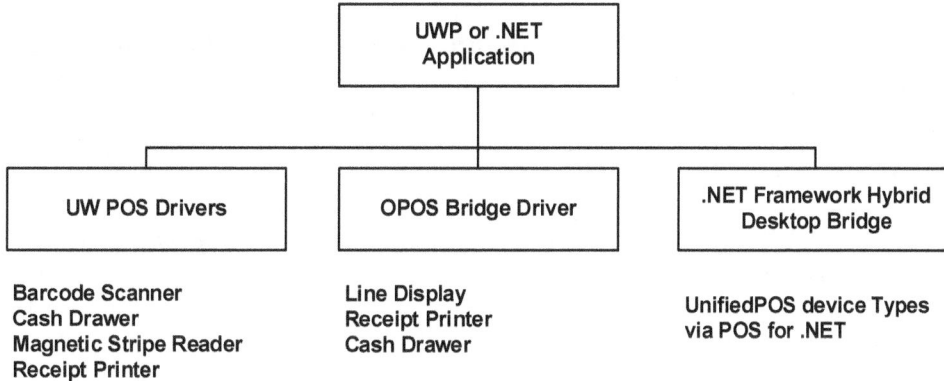

```
                    ┌─────────────────────┐
                    │   UWP or .NET       │
                    │   Application       │
                    └─────────────────────┘
          ┌────────────────┼────────────────────┐
┌─────────────────┐ ┌─────────────────┐ ┌──────────────────────┐
│  UW POS Drivers │ │ OPOS Bridge     │ │ .NET Framework Hybrid│
│                 │ │ Driver          │ │ Desktop Bridge       │
└─────────────────┘ └─────────────────┘ └──────────────────────┘
```

Barcode Scanner
Cash Drawer
Magnetic Stripe Reader
Receipt Printer

Line Display
Receipt Printer
Cash Drawer

UnifiedPOS device Types
via POS for .NET

The assemblies have been tested with POS devices from different POS device manufacturers. There is a hardware support list found here: https://docs.microsoft.com/en-us/windows/uwp/devices-sensors/pos-device-support. For the built-in drivers, class driver support is made available through 5 INF files in the device driver store:

POS Device	INF File	Class GUID
Barcode Scanner	c_barcodescanner.inf	{C243FFBD-3AFC-45E9-B3D3-2BA18BC7EBC5}
Cash Drawer	c_cashdrawer.inf	{772E18F2-8925-4229-A5AC-6453CB482FDA}
Line Display	c_linedisplay.inf	{4FC9541C-0FE6-4480-A4F6-9495A0D17CD2}
Magnetic Stripe Reader (MSR)	c_magneticstripereader.inf	{2A9FE532-0CDC-44F9-9827-76192F2CA2FB}
Receipt Printer	c_receiptprinter.inf	{C7BC9B22-21F0-4F0D-9BB6-66C229B8CD33}

The built-in POS drivers are located in C:\Windows\System32\drivers\UMDF:

Driver	Description
Idtsec.dll	Idtech MSR
mgtdyn.dll	MagTek MSR
Hidscanner.dll	Barcode Scanner
RemotePosDrv.dll	Remote POS Devices Ethernet / Bluetooth
Oposdrv.dll	OPOS Bridge Legacy Driver

The Windows Device Driver Kit (DDK) contains the samples for a magnetic stripe reader and barcode scanner. Windows device driver development is a little more involved than the simple Service Object covered in the last chapter, thus there is no coverage for writing a driver in this book. Writing these device drivers should be done by the POS device manufacturer.

6.3   Exercises 6.1: USB Barcode and Development System Setup

Now that we have covered the new device access sequence and available drivers built into Windows, let's look at the four enumeration options using a barcode scanner. If you are using one of the supported USB barcode scanners that are listed on the Microsoft website, the setup process is as simple as plugging the device into the system.

1. Plug the USB barcode scanner into your Windows development system.
2. Open Device Manager, and you should see the POS HID Barcode scanner listed as one of the devices. If the device shows up as a keyboard, you will have to follow the manufacturer's scan sheet to change to the USB HID mode.

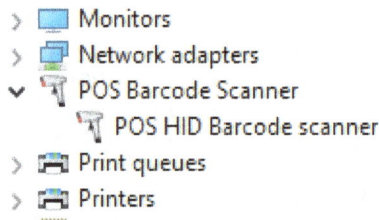

> 🖵 Monitors
> 🖵 Network adapters
> ⌄ 🔫 POS Barcode Scanner
>   🔫 POS HID Barcode scanner
> 🖶 Print queues
> 🖶 Printers

3. Open the properties of the POS HID Barcode scanner.
4. Click on the "Driver" tab, and click on the "Driver Details" button. You will see the HID scanner.dll driver that was discussed in the previous section.

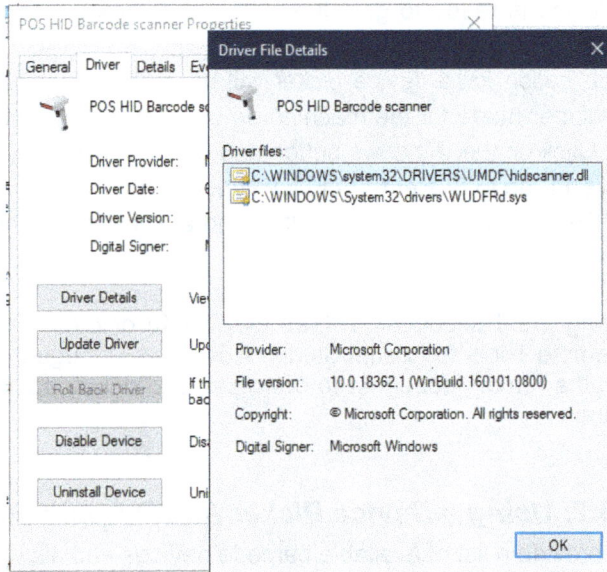

5. Close the Driver Files Details.
6. Click on the "Details" tab, and change the "Property" drop-down to "Class GUID". You will see the same barcode GUID value that is listed in the table from the previous section.

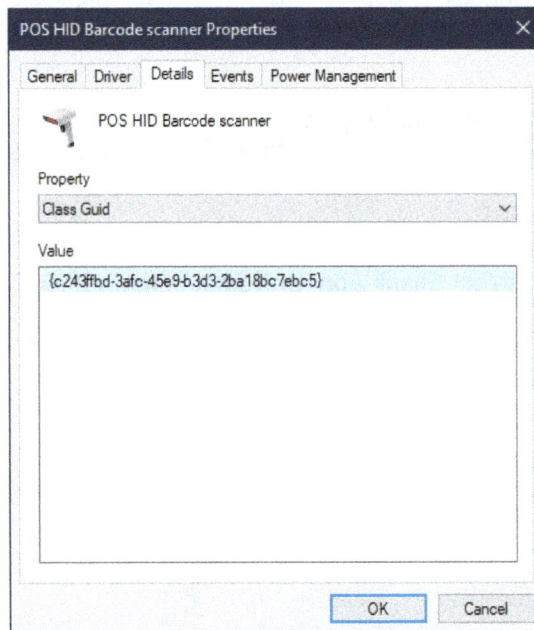

7. The USB device is ready to go. Close the Properties dialog and close Device Manager.
8. The first application for a device picker will be a UWP application. We need to enable Developer mode for the machine on which you are going to run the UWP application. Click on the Windows button and click on "Settings".
9. For Windows 11, go to System->For Developers
10. Select the "Developer mode" radio button and acknowledge any warning dialogs that appear.

Note: If you are using the TabletKiosk TufTab i60XT/i61J or other device as a remote target, install the Remote Tools for Visual Studio. A Remote Debugger tile will appear on the start menu. Run the remote debugger to download and debug applications from your development machine.

6.4 Exercise 6.2: Using a Device Picker (UWP Application)

A device picker will provide a list of available barcode devices and allow the user to select a POS device. The problem is that the device picker is only available for UWP applications and has not been brought over to support .NET. For this exercise, a UWP application will be developed to demonstrate this method. All other applications in this chapter and the next are WPF .NET applications.

Warning: This example is provided as a demonstration of the Device Picker solution. I don't recommend creating UWP applications. WPF .NET applications are the preferred solution for POS device access over UWP applications.

6.4.1 Create the Visual Studio Project
The first step is to create the project and perform the basic setup tasks.

1. Open Visual Studio.
2. Create a new project.
3. Search for and select "Blank App (Universal Windows) for C#" and click Next.

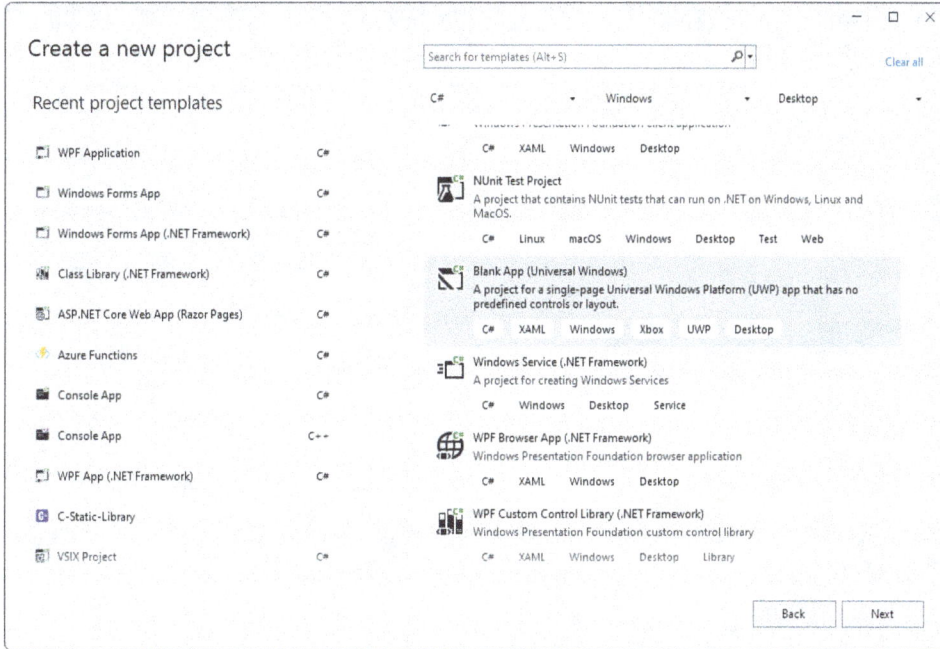

4. Name the project "EX62_BarCode_DevicePicker" and click Create.
5. Just accept the Windows version default options when the dialog appears.
6. Once the project has been created, double-click on the "Package.appxmanifest" file found under the project in the Solution Explorer.
7. Click on "Visual Assets" and use the picture file(s) of your choice to assign pictures to the different tiles and splash screens.
8. Click on "Capabilities".
9. Scroll down the list and check "Point of Service". The Point of Service capability allows access to the Windows.Devices.PointOfService namespace. Failure to select the capability will result in a program crash.

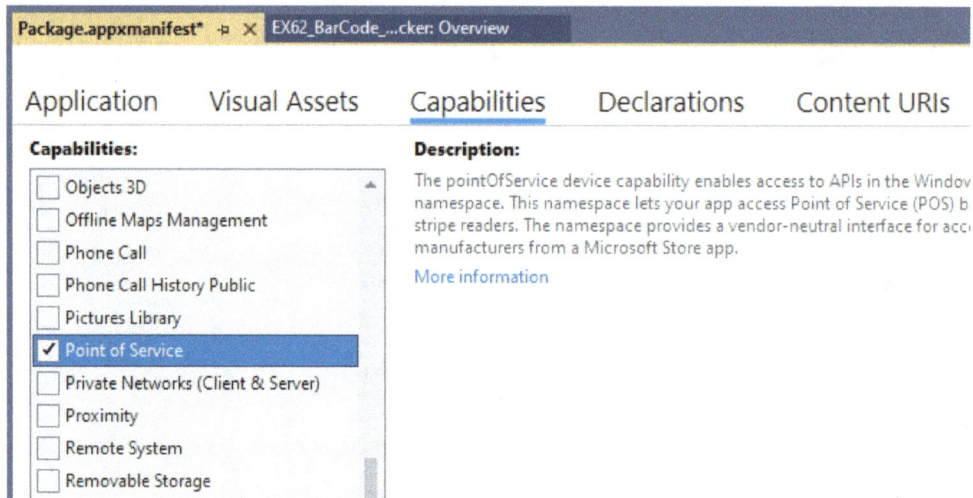

If you were to view the Package.appxmanifest file in code view, you would see the pointOfService capability listed under the Capabilities tag.

```
<Capabilities>
  <DeviceCapability Name="pointOfService" />
</Capabilities>
```

10. Save the project when finished.
11. Close the Package.appxmanifest file.

6.4.2   Set Up the XAML Controls
Now we will set up the UI.

1. In Solution Explorer, open the MainPage.xaml.
2. Add a ListBox under the text block, adjust to allow more controls.
   a. Name: lstItems
3. Add a Button under the list box.
   a. Name: btnShowPicker
   b. Content: Show Device Picker List
4. Add a Text Block at the bottom of the window
   a. Name: ttStatus
   b. Content: Ready

Here is the XAML code:

```xml
<Page
    x:Class="EX62_BarCode_DevicePicker.MainPage"

xmlns="http://schemas.microsoft.com/winfx/2006/xaml/presentatio
n"
    xmlns:x="http://schemas.microsoft.com/winfx/2006/xaml"
    xmlns:local="using:EX62_BarCode_DevicePicker"

xmlns:d="http://schemas.microsoft.com/expression/blend/2008"
    xmlns:mc="http://schemas.openxmlformats.org/markup-
compatibility/2006"
    mc:Ignorable="d"
    Background="{ThemeResource
ApplicationPageBackgroundThemeBrush}">

    <Grid>
        <ListBox x:Name="lstItems" Margin="10,166,-10,394"/>
        <Button    x:Name="btnShowPicker"    Content="Show    Device
Picker    List"    Margin="1047,706,0,0"    VerticalAlignment="Top"
Height="91" Width="325"/>
        <TextBlock x:Name="ttStatus" HorizontalAlignment="Left"
Margin="0,903,0,0"        TextWrapping="Wrap"        Text="Ready"
VerticalAlignment="Top" Height="71" Width="1496"/>

    </Grid>
</Page>
```

The Designer should show the following:

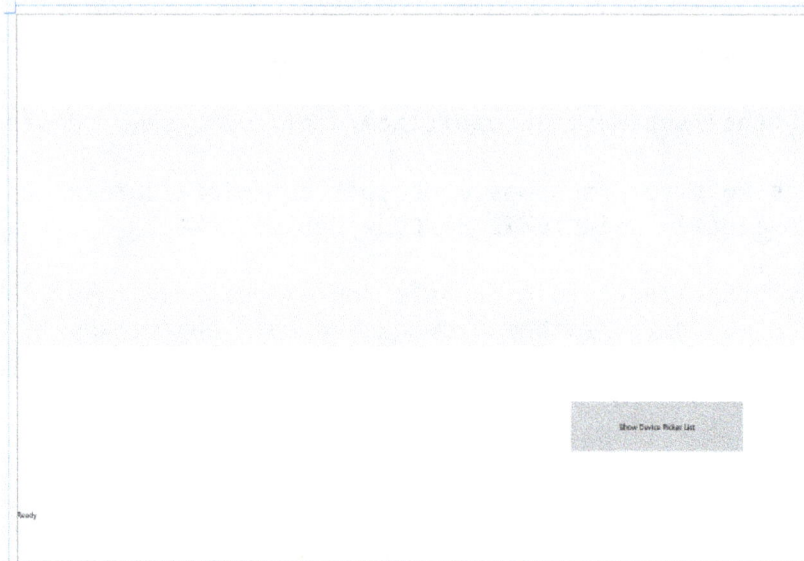

5. Save the MainPage.xaml file.

6.4.3 Write the code

Now, we will put the code behind the interface.

1. Open the MainPage.xaml.cs file.
2. Add the following using statements so we can access the PointOfService namespace, enumeration namespace, and decode the barcode scanner input:

```
using Windows.Devices.PointOfService;
using Windows.Devices.Enumeration;
using Windows.Storage.Streams;
using System.Threading.Tasks;
```

3. Above the main page, add the global properties for a BarcodeScanner and ClaimedBarcodeScanner. These will be used in the methods of the MainPage class.

```
namespace EX62_BarCode_DevicePicker
{
    /// <summary>
    /// Interaction logic for MainWindow.xaml
    /// </summary>
    public partial class MainWindow : Window
    {
```

```
    BarcodeScanner scanner = null;
    ClaimedBarcodeScanner claimedScanner = null;

    public MainWindow()
```

At this point in a POS for .NET application, we would have declared an instance of a POSExplore class and a scanner class. Since there are different methods to enumerate devices, we don't have a POSExplorer class, thus we have to do a little work to connect to the device. The BarcodeScanner class is used to make the connection to the POS device. You could think of the BarcodeScanner class as replacing the POSExplorer class, but the method to enumerate the device will be left up to you. The ClaimedBarcodeScanner class performs a similar operation to the scanner class. ClaimedBarcodeScanner makes the claim to the device, sets up the data received event, and enables the scanner to decode the data from the barcode scanner.

4. After the MainPage() method, add the following Task method, which will set up the DevicePicker so the user can select the barcode device.

```
private async Task<BarcodeScanner> GetBarcodeScanner()
{
    DevicePicker devicePicker = new DevicePicker();

devicePicker.Filter.SupportedDeviceSelectors.Add(BarcodeScanner
.GetDeviceSelector());
    Rect rect = new Rect();
    DeviceInformation        deviceInformation      =      await
devicePicker.PickSingleDeviceAsync(rect);

    //Catch if the user clicks cancel instead of selecting a
device
    try
    {
        scanner                         =                     await
BarcodeScanner.FromIdAsync(deviceInformation.Id);
    }
    catch (NullReferenceException ex)
    {
        ttStatus.Text = "Error getting device: " + ex.ToString();
    }
    return scanner;
}
```

Since the device selection and getting the connection to the device are asynchronous, the asynchronous action, await, is in the async GetBarCodeScanner() method. When a new device picker is instantiated. The device picker instance will call the DeviceSelectors() method to look for any barcode Scanners in the system using an Advanced Query Syntax (AQS) string to get all available scanners in the system. A new rect (rectangle structure) is instantiated, which will hold the list of all barcode devices the device picker finds during the active query. A DeviceInformation object is created with the results of the selection from the device picker. The try-catch attempts to connect to the selected device, and the task returns the Barcode scanner if successful.

5. Below the GetBarcodeScanner() task method, create a new POSSetup() method that will call the GetBarcodeScanner() task and wait for the scanner instance to return. The method will claim and set up the data events for the connected barcode scanner:

```
private async void POSsetup()
{

    scanner = await GetBarcodeScanner();

    if (scanner != null)
    {
        ttStatus.Text    =    "Scanner    Device    Found:    "    +
scanner.DeviceId;
        claimedScanner = await scanner.ClaimScannerAsync();

        if (claimedScanner != null)
        {

            scanner.StatusUpdated += Scanner_StatusUpdated;

            //Avoid other apps from claiming the scanner device
            claimedScanner.ReleaseDeviceRequested              +=
ClaimedScanner_ReleaseDeviceRequested;

            //Setup the data received event handler
            claimedScanner.DataReceived                        +=
ClaimedScanner_DataReceived;

            claimedScanner.IsDecodeDataEnabled = true;
            await claimedScanner.EnableAsync();

            ttStatus.Text += " Claimed";
```

```
        }
        else
        {
            ttStatus.Text = "Claim scanner failed";
        }
    }
    else
    {
        ttStatus.Text += " Scanner not found";
    }
}
```

The first call is to launch the device picker and get the barcode scanner device. Once the connection is made, the following API calls are similar to what is done in a POS for .NET application. An attempt to claim the barcode scanner comes next. If successful, the first thing is to set up an event handler for changes in barcode status. Next comes setting up the event to deny other applications from accessing the barcode scanner while the application is running. Finally, the data received event is set up, enabling the decoding of the data, and makes a call to ready the barcode scanner to receive data events.

6. Enter the following code into the Scanner_StatusUpdate event handler:

```
private async void Scanner_StatusUpdated(BarcodeScanner sender,
BarcodeScannerStatusUpdatedEventArgs args)
{
    uint bcEXStatus = args.ExtendedStatus;
    BarcodeScannerStatus bcStatus = args.Status;


    //Since this event is in a different thread, we need to sync
to the UI thread
    await
Dispatcher.RunAsync(Windows.UI.Core.CoreDispatcherPriority.Norm
al, () =>
    {
        ttStatus.Text = "Status: " + bcStatus.ToString() + " " +
bcEXStatus.ToString();
    });
}
```

This code gets the status of the barcode and sends the results to the status TextBox. The Dispatcher with lambda expression is needed since the event handler is a separate thread from the MainPage thread UI. The dispatcher call is used to marshal the data between the threads. Since the dispatcher is asynchronous, "async" must be added to the Scanner_StatusUpdated() method declaration.

7. Add the following code to the ClaimedScanner_DataReceived() event handler:

```
private                    async                    void
ClaimedScanner_DataReceived(ClaimedBarcodeScanner        sender,
BarcodeScannerDataReceivedEventArgs args)
{
    //Read the data from the buffer and convert to a string.
    var                    scanDataLabelReader            =
DataReader.FromBuffer(args.Report.ScanDataLabel);
    string                 stringDataLabel                =
scanDataLabelReader.ReadString(args.Report.ScanDataLabel.Length
);

    //Since this event is in a different thread, we need to sync
to the UI thread
    await
Dispatcher.RunAsync(Windows.UI.Core.CoreDispatcherPriority.Norm
al, () =>
    {
        lstItems.Items.Add(stringDataLabel);
    });
}
```

The barcode scanner data is read from the buffer first. The args.Report.ScanDataLabel is the decoded data extracted from the raw barcode data. There is no need to decode the full raw barcode data yourself. If you do want to process the full raw barcode data, then you could also retrieve the full raw barcode data with args.Report.ScanData. There is also the ability to retrieve with args.Report.ScanDataType, which supports 94 symbologies. The claimedBarcodeScanner.IsDecodeDataEnabled must be set to "true" to get the barcode data. The data is then converted to a string. The Dispatcher with Lambda expression is needed, again, since the event handler is a separate thread from the MainPage thread UI. As before, remember to add "async" to the method declaration.

8. Add the following code to the ClaimedScanner_ReleaseDeviceRequested event handler:

```
private async void ClaimedScanner_ReleaseDeviceRequested(object
sender, ClaimedBarcodeScanner e)
{
    e.RetainDevice();
```

```
    await
Dispatcher.RunAsync(Windows.UI.Core.CoreDispatcherPriority.Norm
al, () =>
    {
        ttStatus.Text = "Event ReleaseDeviceRequested received.
Retaining the barcode scanner.";
    });
}
```

The code just informs the user that another application wanted to claim the scanner, and the application will continue to retain the device.

9. Save the file.
10. Back in the MainPage.xaml file, double-click on the button to create a button event handler.
11. Go back to MainPage.xaml.cs file and fill in the following code for the btnShowPicker_Click() handler:

```
private void btnShowPicker_Click(object sender, RoutedEventArgs e)
{
    POSsetup();
}
```

The button simply shows the picker again, so the user can select a new device if one is available.

12. In the main method, add the call to the POSSetup() method:

```
public MainPage()
{
    this.InitializeComponent();
    POSsetup();
}
```

13. Remove any unused "using" declarations.
14. Save the whole project.

6.4.4   Test the Application
Now, we are ready to test the application.

1. In Visual Studio, build the application.
2. If there are any errors, correct them, and rebuild the application.
3. Start the debugger.

155

4. Once the application runs, a picker list will fly out from the top, left.

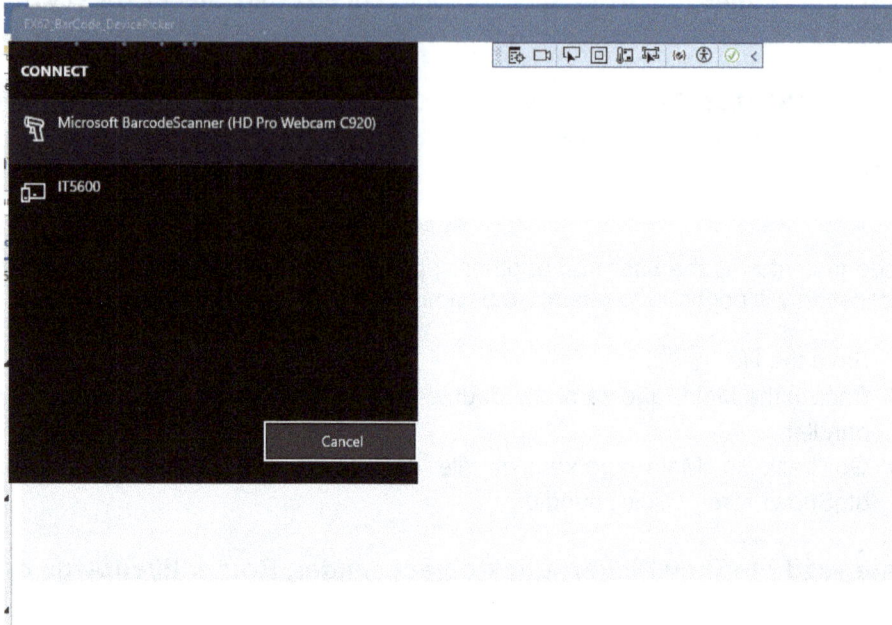

5. Select your barcode scanner. Carefully select the right device as there might be multiple devices that show up as barcodes including web cameras.
6. The status TextBox should show the barcode scanner found and claimed.

Scanner Device Found: \\?\HID#VID_0536&PID_0167#6&2be85541&1&0000#{c243ffbd-3afc-45e9-b3d3-2ba18bc7ebc5}\POSBarcodeScanner Claimed

7. Scan a barcode, and the decoded barcode data results should appear in the ListBox.

853053002176

735858456296

9798985417203

010164089964

8. Stop debugging.
9. Change the ClaimedScanner_DataReceived method as follows to get the full raw barcode data and the barcode symbology type:

```
private                        async                        void
ClaimedScanner_DataReceived(ClaimedBarcodeScanner           sender,
BarcodeScannerDataReceivedEventArgs args)
{
    //Read the data from the buffer and convert to a string.
    var                  scanDataLabelReader                     =
DataReader.FromBuffer(args.Report.ScanDataLabel);
    string                   stringDataLabel                     =
scanDataLabelReader.ReadString(args.Report.ScanDataLabel.Length
);

    var                      scanDataReader                      =
DataReader.FromBuffer(args.Report.ScanData);
    string                     stringData                        =
scanDataReader.ReadString(args.Report.ScanData.Length);
    string stringDataType = args.Report.ScanDataType.ToString();



    //Since this event is in a different thread, we need to sync
to the UI thread
    await
Dispatcher.RunAsync(Windows.UI.Core.CoreDispatcherPriority.Norm
al, () =>
```

157

```
    {
        lstItems.Items.Add(stringDataLabel);
        lstItems.Items.Add(stringData);
        lstItems.Items.Add(stringDataType);


    });
}
```

10. Run the debugger again, and scan a barcode. This time you will get three items in the display: decoded data, full raw data, and the barcode symbology type.

853053002176

JE0853053002176c

107

6.5 Exercise 6.3: Remembering a device ID

Why would you want the user to select the same barcode from the device picker list each time the application starts? Wouldn't the DefaultAsync be simpler to implement? A system might have multiple devices listed as barcode scanners, and the scanner set as the default can result in an unknown device getting selected when the application starts. For the Device Picker, you could actually save the device ID so that the application can retrieve the saved device ID on the next session of the application and simply connect to the saved device without showing the picker. The Windows.Storage namespace provides the mechanism to save data between application sessions. In this exercise, we will modify Exercise 6.2: Using a Device Picker to add the ability to save settings locally.

Warning: This information is provided for completeness of the Device Picker method. WPF .NET applications are the preferred solution for POS device access over UWP applications.

Note: This solution only works for UWP applications. For .NET and .NET Framework, you can use the registry to store values in HKCU. Chapter 9 will demonstrate a solution for .NET and .NET Framework.

1. Open the Barcode-DevicePicker project in Visual Studio.

2. In the using section add the Windows.Storage namespace:

```
using Windows.Devices.PointOfService;
using Windows.Devices.Enumeration;
using Windows.Storage.Streams;
using System.Threading.Tasks;
using Windows.Storage;
```

3. After the global properties for BarcodeScanner and ClaimedBarcodeScanner, add the ApplicationDataContainer property:

```
public sealed partial class MainPage : Page
{

    BarcodeScanner scanner = null;
    ClaimedBarcodeScanner claimedScanner = null;
    ApplicationDataContainer localSettings = null;
```

4. In the MainPage() method, add the instantiation of the localSettings.

```
public MainPage()
{
    this.InitializeComponent();
    localSettings = ApplicationData.Current.LocalSettings;
    POSsetup();
}
```

The device ID is a string that looks similar to the following:

```
\\?\HID#VID_0536&PID_0167#7&2be85541&0&0000#{c243ffbd-3afc-45e9-
b3d3-2ba18bc7ebc5}\POSBarcodeScanner
```

It contains the device's instance path, the POS device Class ID, and the POS class name.

5. We want to store the string between reboots. In the GetBarcodeScanner(), within the Try of the try-catch add the following code to store the device ID.

    localSettings.Values["deviceID"] = deviceInformation.Id;

This will store the device ID in a value setting called device ID after the user has selected the device from the device picker. The whole method should look as follows:

```csharp
private async Task<BarcodeScanner> GetBarcodeScanner()
{
    DevicePicker devicePicker = new DevicePicker();

devicePicker.Filter.SupportedDeviceSelectors.Add(BarcodeScanner
.GetDeviceSelector());
    Rect rect = new Rect();
    DeviceInformation        deviceInformation       =        await
devicePicker.PickSingleDeviceAsync(rect);

    //Catch if the user clicks cancel instead of selecting a
device
    try
    {
        scanner                          =                      await
BarcodeScanner.FromIdAsync(deviceInformation.Id);
        localSettings.Values["deviceID"] = deviceInformation.Id;
    }
    catch (NullReferenceException ex)
    {
        ttStatus.Text = "Error getting device: " + ex.ToString();
    }
    return scanner;
}
```

6. In the POSSetup() method, we will now add the logic to determine if the device ID exists or not. If the device ID doesn't exist, the device picker will be shown to the user. If there is a device ID, the connection using the device ID will be made immediately without showing the device picker. Here is the logic to add:

```csharp
private async void POSsetup()
{

    if (localSettings.Values["deviceID"] != null)
    {
        scanner                         =                       await
BarcodeScanner.FromIdAsync(localSettings.Values["deviceID"].ToS
tring());
    }
    else
    {
        scanner = await GetBarcodeScanner();
```

```
    }

    if (scanner != null)
    {
        ttStatus.Text   =   "Scanner   Device   Found:   "   +
scanner.DeviceId;
        claimedScanner = await scanner.ClaimScannerAsync();

        if (claimedScanner != null)
        {

            scanner.StatusUpdated += Scanner_StatusUpdated;

            //Avoid other apps from claiming the scanner device
            claimedScanner.ReleaseDeviceRequested           +=
ClaimedScanner_ReleaseDeviceRequested;

            //Setup the data received event handler
            claimedScanner.DataReceived                     +=
ClaimedScanner_DataReceived;

            claimedScanner.IsDecodeDataEnabled = true;
            await claimedScanner.EnableAsync();

            ttStatus.Text += " Claimed";
        }
        else
        {
            ttStatus.Text = "Claim scanner failed";
        }
    }
    else
    {
        ttStatus.Text += " Scanner not found";
    }
}
```

7. Finally, we will modify the btnShowPicker_Click event handler to clear out the device ID value so the device picker can be shown again:

```
private void btnShowPicker_Click(object sender, RoutedEventArgs e)
{
    localSettings.Values["deviceID"] = null;
    POSsetup();
}
```

8. Start debugging the application.
9. The device picker will show up the first time. Select the barcode scanner and perform a scan.
10. Stop the debug session, which closes the application.
11. Start debugging the application again. This time you don't get a device picker. The application will use the previous device ID value and connect to the device automatically.
12. Click on the "Show Device Picker List button", clear the device ID value, and show the device picker.
13. Stop debugging when finished.
14. Since this is a UWP application, uninstall the application from the start menu.

The ability to store the device ID locally makes the user experience a little less taxing each time the application starts. Between the Device Picker enumeration method and DefaultAsync() method, the Device Picker with setting storage is a better solution. Also, DefaultAsync is only for locally connected devices, which doesn't work well for devices on the network or connected via Bluetooth. Unfortunately, Device Picker is only for UWP applications. Let's move on to a .NET application using the aforementioned DefaultAsync() enumeration.

6.6 Exercise 6.4: Getting First Available Device using DefaultAsync()

Most of the time you will have only one POS device connected to the system, in which case, you will not need to gather all the device types. Instead, you can simply get the default device in the system. In this exercise, the application will simply call the DefaultAsync() method to enumerate the barcode scanner.

6.6.1 Create the Visual Studio Project

The first step is to create the project and perform the basic setup tasks.

1. Open Visual Studio.
2. Create a new project.
3. Search for and select "WPF Application for C#" and click Next.

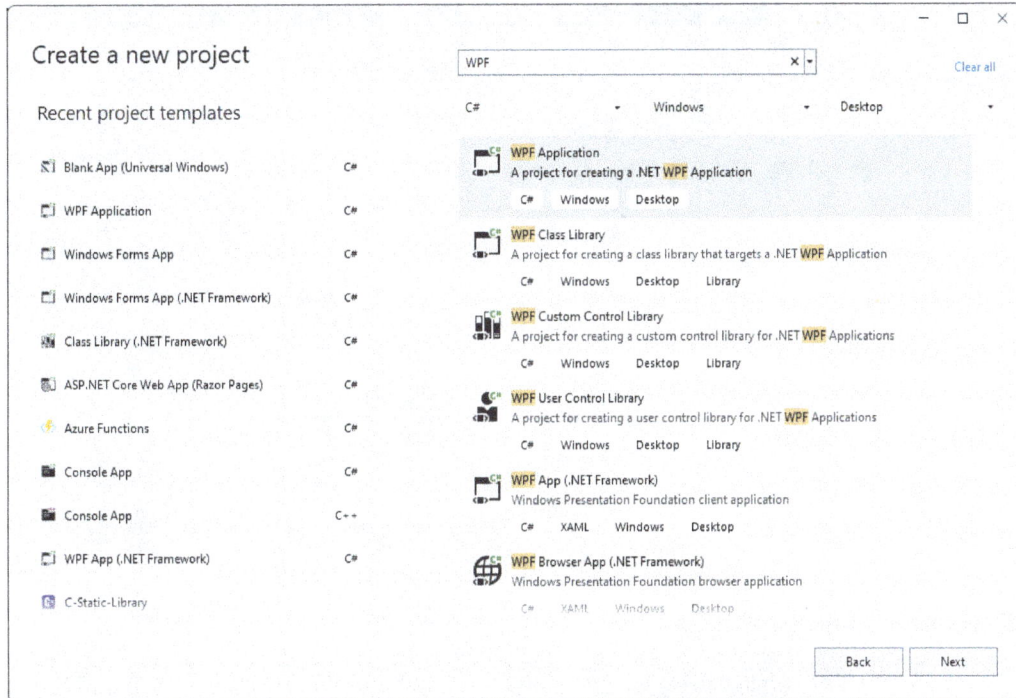

4. Name the project EX64_BarCode_DefaultAsync and click Next.
5. Keep the .NET version at ".NET 8.0 (Long Term Support)" and click Create.
6. The next and most important step: from the menu select Project->
   EX64_BarCode_DefaultAsync properties.
7. Under General, change the following:

   Target OS Version: 10.0.26100.0
   Supported OS Version: 10.0.19041.0

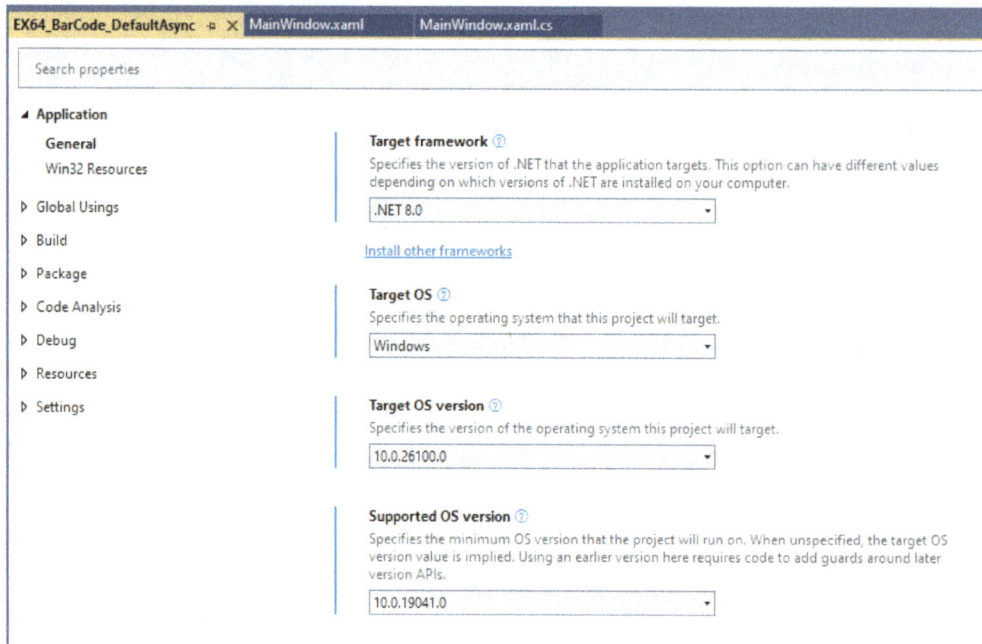

The reason the OS version is important is that the minimum OS version of 10.0.19041.0 is the OS version that supports the POS WinRT namespaces. If you use any OS version below the minimum, you will not be able to get access to the POS WinRT namespaces.

8. Save the project when finished.
9. Close the EX64_BarCode_DefaultAsync properties page.

6.6.2   Set Up the XAML Controls
Now, we will set up the UI.

1. In Solution Explorer, open the MainWindow.xaml.
2. Add a ListBox, and resize to take the upper 2/3 of the page.
   a. Name: lstItems
   b. Font Size 14
3. Add a StatusBar under the ListBox.
   a. Name: sBar
4. Add a TextBlock in the top of the StatusBar.
   a. Name: txtStatus
   b. Font size 11
   c. Text: Ready

Here is the XAML code:

```
<Window x:Class="EX64_BarCode_DefaultAsync.MainWindow"
```

164

```xml
xmlns="http://schemas.microsoft.com/winfx/2006/xaml/presentatio
n"
        xmlns:x="http://schemas.microsoft.com/winfx/2006/xaml"

xmlns:d="http://schemas.microsoft.com/expression/blend/2008"
        xmlns:mc="http://schemas.openxmlformats.org/markup-
compatibility/2006"
        xmlns:local="clr-namespace:EX64_BarCode_DefaultAsync"
        mc:Ignorable="d"
        Title="MainWindow" Height="450" Width="800">
    <Grid>
        <ListBox x:Name="lstItems" d:ItemsSource="{d:SampleData
ItemCount=5}" Margin="0,46,0,143" FontSize="14"/>
        <StatusBar          x:Name="sBar"          Margin="0,387,0,0"
FontSize="11">
            <TextBlock   x:Name="txtStatus"   TextWrapping="Wrap"
Text="Ready" Width="779"/>
        </StatusBar>

    </Grid>
</Window>
```

The Designer should look as follows:

5. Save the MainWindow.xaml file.

6.6.3 Write the code
Now we will put the code behind the interface.

1. Open the MainWindow.xaml.cs file.
2. Add the following using statements so we can access the PointOfService namespace :

```
using Windows.Devices.PointOfService;
using Windows.Storage.Streams;
```

3. Above the main page, add the global properties for a BarcodeScanner and ClaimedBarcodeScanner. These will be used in the methods of the MainPage class.

```
namespace EX64_BarCode_DefaultAsync
{
    /// <summary>
    /// Interaction logic for MainWindow.xaml
    /// </summary>
    ///
    public partial class MainWindow : Window
    {

        BarcodeScanner scanner = null;
        ClaimedBarcodeScanner claimedScanner = null;

        public MainWindow()
        {
            InitializeComponent();
        }
```

4. After the MainWindow() method, create a POSSetup() method. As you hit the +=, hit the tab to generate the event handler.

```
private async void POSsetup()
{
    scanner = await BarcodeScanner.GetDefaultAsync();

    if (scanner != null)
    {
```

```
            txtStatus.Text   =   "Scanner   Device   Found:   "   +
scanner.DeviceId;
        claimedScanner = await scanner.ClaimScannerAsync();
        if (claimedScanner != null)
        {
            scanner.StatusUpdated += Scanner_StatusUpdated;

            //Avoid other apps from claiming the scanner device
            claimedScanner.ReleaseDeviceRequested           +=
ClaimedScanner_ReleaseDeviceRequested;

            //Setup the data received event handler
            claimedScanner.DataReceived                     +=
ClaimedScanner_DataReceived;

            claimedScanner.IsDecodeDataEnabled = true;
            await claimedScanner.EnableAsync();
            txtStatus.Text += " Claimed";

        }
        else
        {
            txtStatus.Text = "Claim scanner failed";
        }
    }
    else
    {
        txtStatus.Text += " Scanner not found";
    }
}
```

There is no picker task method required for a GetDefaultAsync() call. The call to get the default barcode device simply returns the first available barcode reader. The rest of the code is the same as before.

5. Add the code for the ClaimedScanner_DataReceived event handler method:

```
private  void  ClaimedScanner_DataReceived(ClaimedBarcodeScanner
sender, BarcodeScannerDataReceivedEventArgs args)
{
    //Read the data from the buffer and convert to a string.
```

```
    var              scanDataLabelReader              =
DataReader.FromBuffer(args.Report.ScanDataLabel);
    string              stringDataLabel              =
scanDataLabelReader.ReadString(args.Report.ScanDataLabel.Length
);

    //Since this event is in a different thread, we need to sync
to the UI thread
    Dispatcher.BeginInvoke((Action)(()                =>
lstItems.Items.Add(stringDataLabel)));
    Dispatcher.BeginInvoke((Action)(()   =>   txtStatus.Text   =
"Ready"));
}
```

6. Add the code for the ClaimedScanner_ReleaseDeviceRequested event handler method:

```
private    void    ClaimedScanner_ReleaseDeviceRequested(object?
sender, ClaimedBarcodeScanner e)
{
    e.RetainDevice();

    Dispatcher.BeginInvoke((Action)(()   =>   txtStatus.Text   =
"Event ReleaseDeviceRequested received. Retaining the barcode
scanner."));
}
```

7. Add the code for the Scanner_StatusUpdated event handler method:

```
private    void    Scanner_StatusUpdated(BarcodeScanner    sender,
BarcodeScannerStatusUpdatedEventArgs args)
{
    uint bcEXStatus = args.ExtendedStatus;
    BarcodeScannerStatus bcStatus = args.Status;
    //Since this event is in a different thread, we need to sync
to the UI thread
    Dispatcher.BeginInvoke((Action)(()   =>   txtStatus.Text   =
"Status: " + bcStatus.ToString() + " " + bcEXStatus.ToString()));
}
```

8. Add POSSetup() to the MainWindow() method:

```
public MainWindow()
```

168

```
{
    InitializeComponent();
    POSsetup();
}
```

9. Save the project.

6.6.4 Test the Application

Now, we are ready to test the application.

1. In Visual Studio, build the application.
2. If there are any errors, correct them, and rebuild the application.
3. Plug the barcode scanner into your development machine, and run the debugger locally.

The application will start up. The barcode scanner should be found and claimed.

4. Scan some barcodes and the results will appear in the list box.

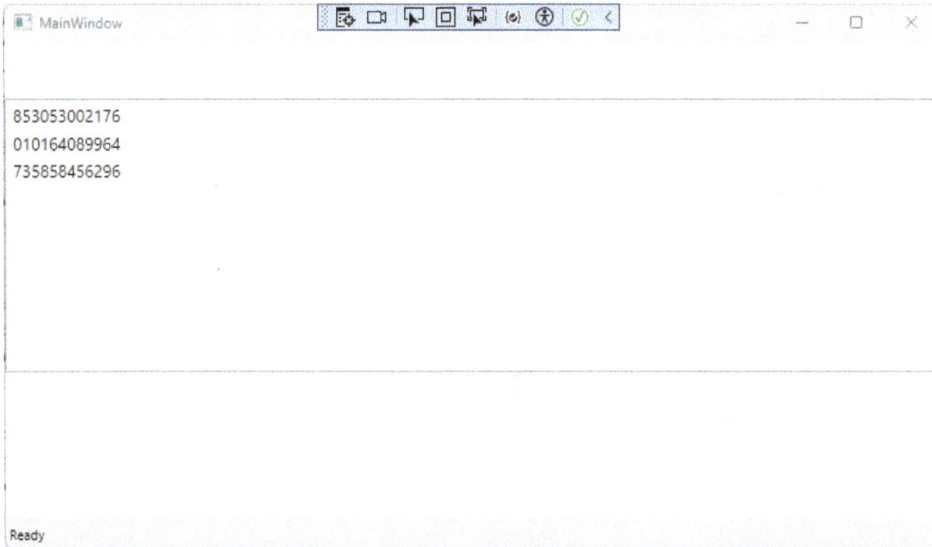

Even though there is a webcam in my system, the USB barcode scanner comes up first and is claimed by the application. If the barcode scanner was not attached, a different GLOBAL device is clammed, but it is not called out as a POSBarcodeScanner.

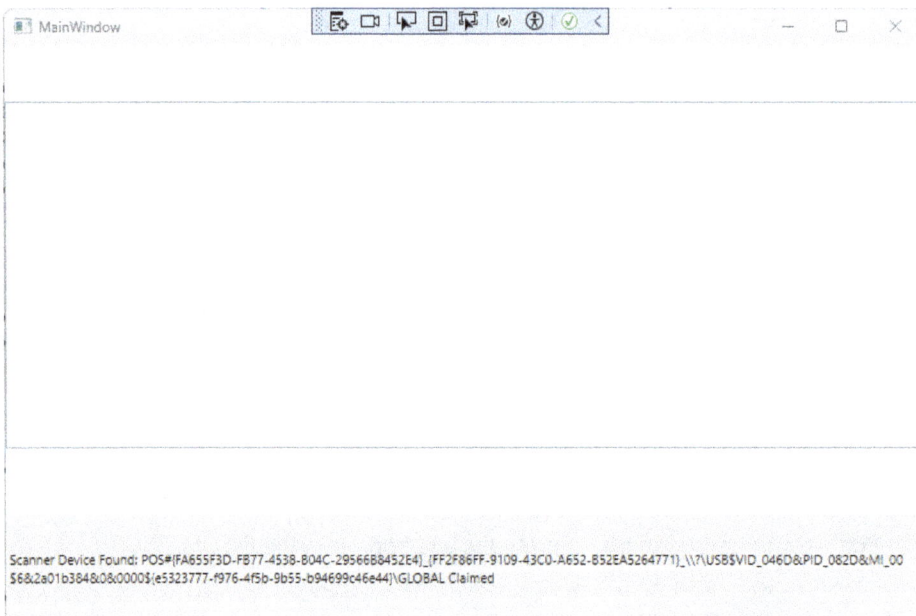

The application is not usable without a barcode scanner, and plugging in one after the application starts is not going to work. The next two exercises look at the last two enumeration methods.

6.7 Exercise 6.5: Snapshot of Devices for Custom Interfaces

The Device Picker enumeration solution is not available for .NET applications, but one could create a custom interface for listing all the devices or a filtered set of devices. This is known as the Snapshot enumeration method. The Snapshot enumeration method uses the DeviceIntormationCollection class to gather a list of POS device types. The collection can be displayed in a control of your choosing such as ListBox, data grid, etc. For this exercise, we will implement the Snapshot enumeration method using a ListBox control to list the available barcode scanners.

6.7.1 Create the Visual Studio Project

The first step is to create the project and perform the basic setup tasks.

1. Open Visual Studio.
2. Create a new project.
3. Search for and select "WPF Application for C#" and click Next.

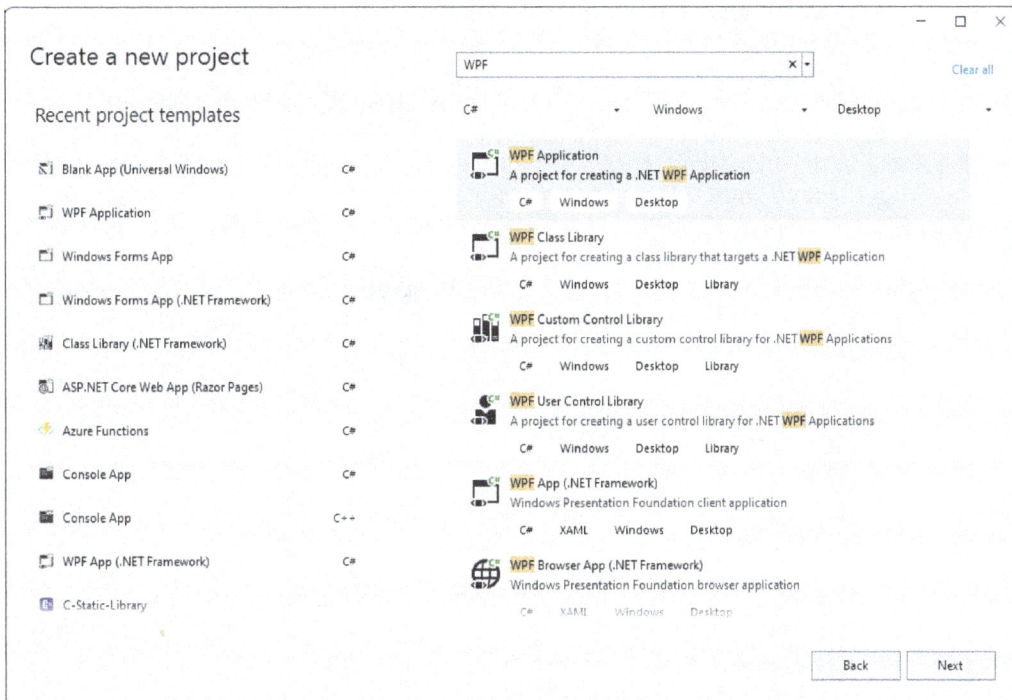

4. Name the project EX65_BarCode_Snapshot and click Next.
5. Keep the .NET version at ".NET 8.0 (Long Term Support)" and click Create.

171

6. The next and most important step: from the menu select Project->
   EX64_BarCode_DefaultAsync properties.
7. Under General, change the following:

   Target OS Version: 10.0.26100.0
   Supported OS Version: 10.0.19041.0

The reason the OS version is important is that the minimum OS version of 10.0.19041.0 is the OS version that supports the POS WinRT namespaces. If you use any OS version below the minimum, you will not be able to get access to the POS WinRT namespaces.

8. Save the project when finished.
9. Close the EX65_BarCode_Snapshot properties page.

6.7.2   Set Up the XAML Controls
Now we will set up the UI.

1. In Solution Explorer, open the MainWindow.xaml.
2. Add a Label in the top, left corner.
   a.   Name: lblScandItems
   b.   Font size 14 Bold
   c.   Text: Scanned Items
3. Add a ListBox under the label and resize the LisBox to fill ¼ of the page.
   a.   Name: lstItems
4. Add a Label under the ListBox
   a.   Name: lblPOSDevces
   b.   Font size 14 Bold
   c.   Text: Barcode Scanner
5. Add a ListBox under the label and resize to go under the label and leave enough room at the bottom for the status strip.
   a.   Name: lstDevices
6. Add a StatusBar
   a.   Name: sBAR
7. Add a TextBlock to the StatusBar .
   a.   Name: txtStatus

Here is the XAML code:

```
<Window x:Class="EX65_BarCode_Snapshot.MainWindow"

xmlns="http://schemas.microsoft.com/winfx/2006/xaml/presentatio
n"
        xmlns:x="http://schemas.microsoft.com/winfx/2006/xaml"

xmlns:d="http://schemas.microsoft.com/expression/blend/2008"
```

```
        xmlns:mc="http://schemas.openxmlformats.org/markup-
compatibility/2006"
        xmlns:local="clr-namespace:EX65_BarCode_Snapshot"
        mc:Ignorable="d"
        Title="MainWindow" Height="450" Width="800">
    <Grid>
        <ListBox x:Name="lstItems" d:ItemsSource="{d:SampleData
ItemCount=5}" Margin="0,36,0,195"/>
        <Label x:Name="lblScandItems" Content="Scanned Items:"
HorizontalAlignment="Left"                    Margin="10,5,0,0"
VerticalAlignment="Top" FontWeight="Bold" FontSize="14"/>
        <ListBox                              x:Name="lstDevices"
d:ItemsSource="{d:SampleData  ItemCount=5}"  Margin="0,270,0,63"
SelectionChanged="lstDevices_SelectionChanged"/>
        <Label x:Name="lblPOSDevces" Content="Barcode Scanners"
HorizontalAlignment="Left"                  Margin="10,244,0,0"
VerticalAlignment="Top" FontWeight="Bold" FontSize="14"/>
        <StatusBar x:Name="sBar" Margin="0,371,0,0">
            <TextBlock   x:Name="txtStatus"   TextWrapping="Wrap"
Text="Ready" Height="49" Width="792"/>
        </StatusBar>

    </Grid>
</Window>
```

Here is what the Designer should look like:

8.   Save the MainWindow.xaml file.

6.7.3   Write the code
Now we will put the code behind the interface.

1.   Open the MainWindow.xaml.cs file.
2.   Add the following using statements so we can access the PointOfService namespace and the namespaces to enumerate the POS devices and decode the barcode scanner input:

```
using Windows.Devices.PointOfService;
using Windows.Devices.Enumeration;
using Windows.Security.Cryptography;
using System.Collections;
using Windows.Storage.Streams;
```

3.   Above the main page, add the global properties for a BarcodeScanner and ClaimedBarcodeScanner. Also, instantiate a new Arraylist that will collect all the barcode scanner device IDs. These will be used in the methods of the MainPage class.

```
public partial class MainWindow : Window
{
    BarcodeScanner scanner = null;
```

```
ClaimedBarcodeScanner claimedScanner = null;
ArrayList myDeviceIDs = new ArrayList();

public MainWindow()
{
```

4. After the MainWindow() method create a POSSetup() method and add the following code:

```
private async void POSSetup()
{
    DeviceInformationCollection    deviceCollection    =    await
DeviceInformation.FindAllAsync(BarcodeScanner.GetDeviceSelector
(PosConnectionTypes.All));

    foreach (DeviceInformation devinfo in deviceCollection)
    {
        lstDevices.Items.Add(devinfo.Name);
        myDeviceIDs.Add(devinfo.Id);
    }
}
```

An instance of the DeviceInformationCollection is used to gather all the barcode scanners attached to the system. The foreach loop will then add the device name to the device ListBox and the device ID to the ArrayList. Once all the devices have been added to the list, the user will tap on the desired device to select it. We now need to add a tapped event handler.

5. In the MainPage.xml, go to the XAML code.
6. In the lstDevice ListBox, enter SelectionChange= and hit the tab key twice to generate the event handler in MainPage.xaml.cs.

```
<ListBox       x:Name="lstDevices"       d:ItemsSource="{d:SampleData
ItemCount=5}"                              Margin="0,270,0,63"
SelectionChanged="lstDevices_SelectionChanged"/>
```

7. The lstDevices_SelectionChanged handler will perform the device connection and setup of the barcode handers. Enter the following code in the lstDevices_SelectionChanged handler:

```
private async void lstDevices_SelectionChanged(object sender,
SelectionChangedEventArgs e)
{
```

```
    ListBox? lb = sender as ListBox;
    if (lb != null)
    {
        txtStatus.Text = lb.SelectedItem.ToString();
        scanner                        =                  await
BarcodeScanner.FromIdAsync((string)myDeviceIDs[lb.SelectedIndex
]);

        if (scanner != null)
        {
            txtStatus.Text = "Scanner Device selected: " +
scanner.DeviceId;
            claimedScanner = await scanner.ClaimScannerAsync();
            if (claimedScanner != null)
            {
                scanner.StatusUpdated += Scanner_StatusUpdated;
                claimedScanner.ReleaseDeviceRequested        +=
ClaimedScanner_ReleaseDeviceRequested;
                claimedScanner.DataReceived                  +=
ClaimedScanner_DataReceived;
                claimedScanner.IsDecodeDataEnabled = true;
                await claimedScanner.EnableAsync();
                txtStatus.Text += " Claimed";
            }
            else
            {
                txtStatus.Text = "Claim scanner failed";
            }
        }
        else
        {
            txtStatus.Text = "Scanner Not Found";
        }
    }
    else
    {
        txtStatus.Text = "List item failed";
    }
}
```

The ListBox item's selection index is used to get the device ID from the ArrayList. The rest of the code is the same as the device picker exercise.

8. Add the code for the ClaimedScanner_DataReceived event handler method:

```
private void ClaimedScanner_DataReceived(ClaimedBarcodeScanner
sender, BarcodeScannerDataReceivedEventArgs args)
{
    //Read the data from the buffer and convert to a string.
    var                    scanDataLabelReader                    =
DataReader.FromBuffer(args.Report.ScanDataLabel);
    string                 stringDataLabel                        =
scanDataLabelReader.ReadString(args.Report.ScanDataLabel.Length
);

    //Since this event is in a different thread, we need to sync
to the UI thread
    Dispatcher.BeginInvoke((Action)(()                          =>
lstItems.Items.Add(stringDataLabel)));
    Dispatcher.BeginInvoke((Action)(()   =>   txtStatus.Text   =
"Ready"));
}
```

9. Add the code for the ClaimedScanner_ReleaseDeviceRequested event handler method:

```
private   void   ClaimedScanner_ReleaseDeviceRequested(object?
sender, ClaimedBarcodeScanner e)
{
    e.RetainDevice();

    Dispatcher.BeginInvoke((Action)(()   =>   txtStatus.Text   =
"Event ReleaseDeviceRequested received. Retaining the barcode
scanner."));
}
```

10. Add the code for the Scanner_StatusUpdated event handler method:

```
private   void   Scanner_StatusUpdated(BarcodeScanner   sender,
BarcodeScannerStatusUpdatedEventArgs args)
{
    uint bcEXStatus = args.ExtendedStatus;
    BarcodeScannerStatus bcStatus = args.Status;
    //Since this event is in a different thread, we need to sync
to the UI thread
```

177

```
    Dispatcher.BeginInvoke((Action)(()    =>    txtStatus.Text    =
"Status: " + bcStatus.ToString() + " " + bcEXStatus.ToString()));
}
```

11. Add POSSetup() to the MainWindow() method:

```
public MainWindow()
{
    InitializeComponent();
    POSsetup();
}
```

12. Save the project.

6.7.4   Test the Application
Now, we are ready to test the application.

1. In Visual Studio, build the application.
2. If there are any errors, correct them, and rebuild the application.
3. Plug the barcode scanner into your development machine, and run the debugger locally. The application will start up, and it will take a little time to list all available barcode scanners in the ListBox.

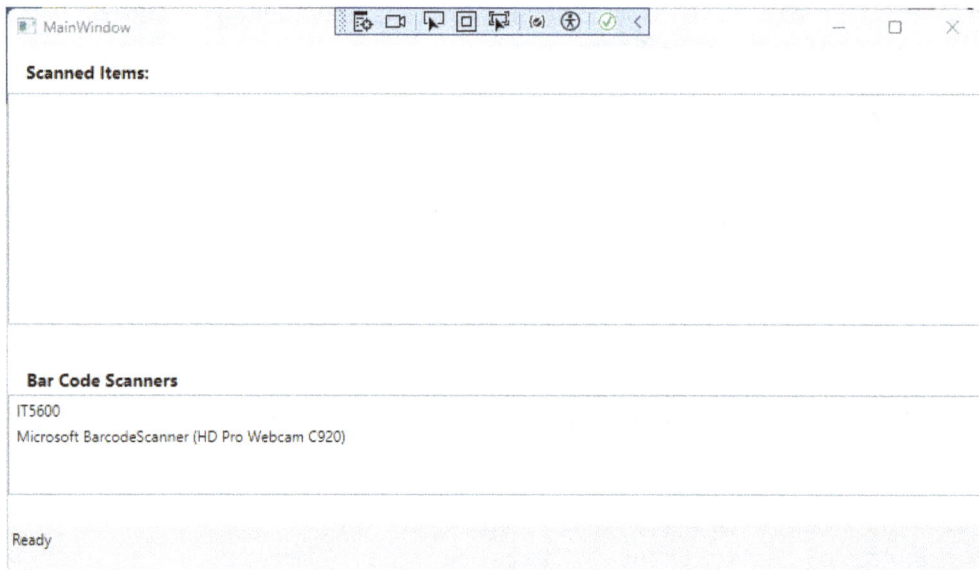

4. Tap on the barcode scanner in the list. The application should make the connection and claim the device.

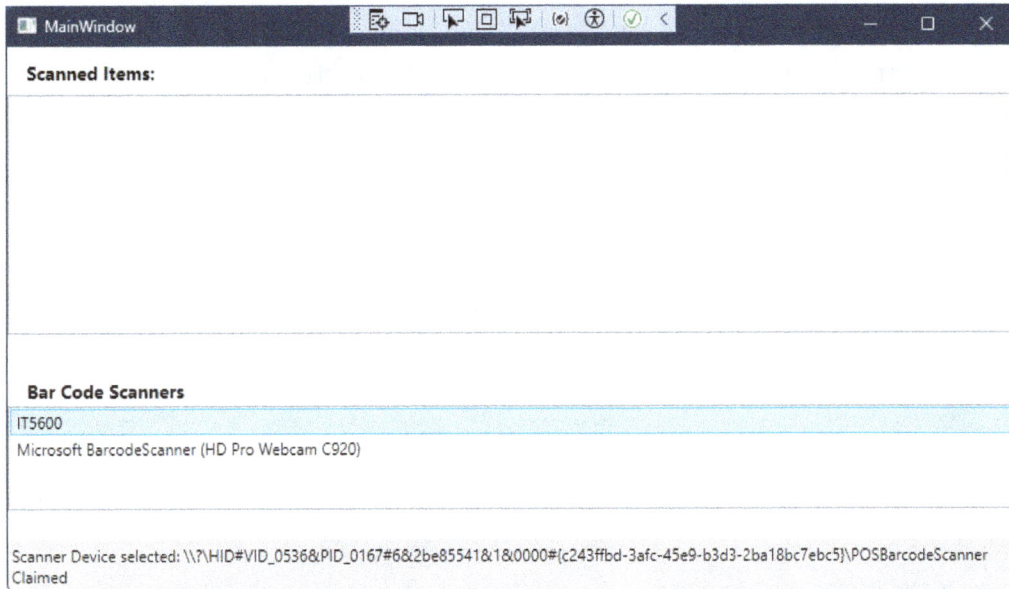

5. Scan a barcode and the output data should be displayed in the lstItems ListBox.
6. In the picture above, there is the HHP IT5600 board code scanner and the WebCam. If you have something similar, test selecting the webcam, scan a barcode with the WebCam scanner, and switch back to the barcode scanner. When you switch to the Webcam, the other device will display the change event, but the barcode will still work.
7. Stop the debugger when finished.

6.8 *Exercise 6.6: Using a Device Watcher*

For POS for .NET applications, USB barcode and MSR devices supported device added and removed events with the POS for .NET runtime managing the event signaling. The first 3 device enumerations don't really have the same dynamic device handling as POS for .NET, which is fine as not every POS device is a USB Plug and Play device. If the system you are developing for is going to be using USB Plug and Play devices and it is a good possibility these devices are going to be removed or replaced, the last device enumeration solution, device watcher, is the solution. In this final chapter exercise, we will see how device watcher aligns with the POS for .NET device added and removed events.

6.8.1   Create the Visual Studio Project

The first step is to create the project and perform the basic setup tasks.

1. Open Visual Studio.
2. Create a new project.
3. Search for and select "WPF Application for C#" and click Next.

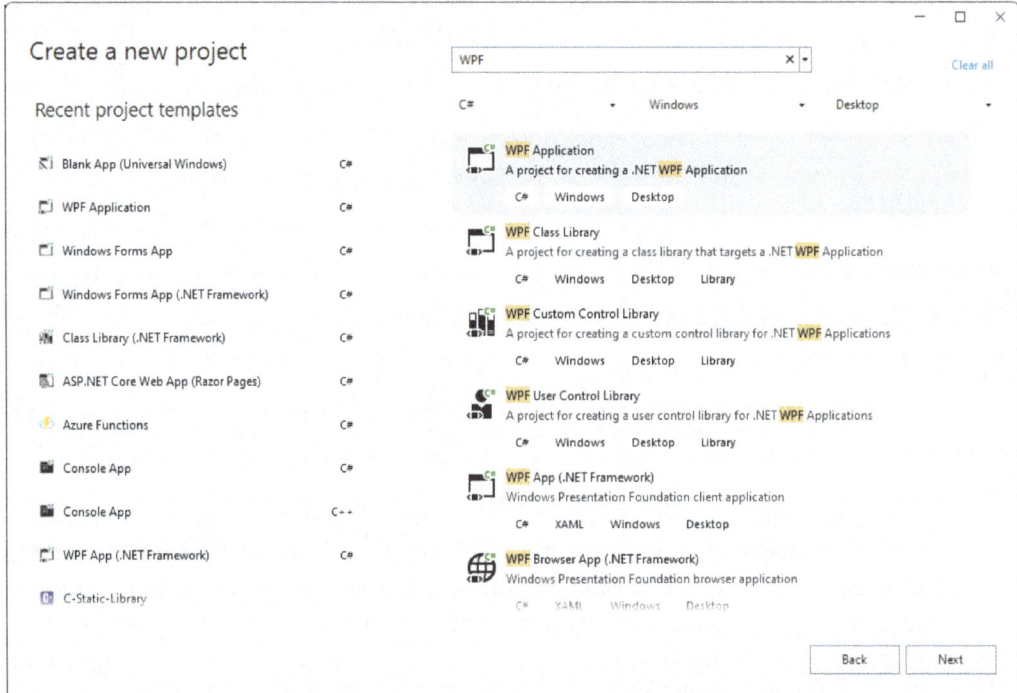

4. Name the project EX66_BarCode_DeviceWatcher and click Next.
5. Keep the .NET version at ".NET 8.0 (Long Term Support)" and click Create.
6. The next and most important step: from the menu select Project-> EX64_BarCode_DefaultAsync properties.
7. Under General, change the following:

> Target OS Version: 10.0.26100.0
> Supported OS Version: 10.0.19041.0

The reason the OS version is important is that the minimum OS version of 10.0.19041.0 is the version that supports the POS WinRT namespaces. If you use any OS version below the minimum, you will not be able to get access to the POS WinRT namespaces.

8. Save the project when finished.
9. Close the EX66_BarCode_DeviceWatcher properties page.

180

6.8.2   Set Up the XAML Controls
Now, we will set up the UI.

1. In Solution Explorer, open the MainWindow.xaml.
2. Add a Label in the top, left corner.
   a. Name: lblScandItems
   b. Font size 14 Bold
   c. Text: Scanned Items
3. Add a ListBox under the label and resize the LisBox to fill 3/4 of the page.
   a. Name: lstItems
4. Add a StatusBar at the bottom of the page.
   a. Name: sBAR
5. Add a TextBlock to the StatusBar.
   a. Name: txtStatus

Here is the XAML code:

```
<Window x:Class="EX66_BarC0de_DeviceWatcher.MainWindow"

xmlns="http://schemas.microsoft.com/winfx/2006/xaml/presentatio
n"
        xmlns:x="http://schemas.microsoft.com/winfx/2006/xaml"

xmlns:d="http://schemas.microsoft.com/expression/blend/2008"
        xmlns:mc="http://schemas.openxmlformats.org/markup-
compatibility/2006"
        xmlns:local="clr-namespace:EX66_BarC0de_DeviceWatcher"
        mc:Ignorable="d"
        Title="MainWindow" Height="450" Width="800">
    <Grid>
        <ListBox x:Name="lstItems" d:ItemsSource="{d:SampleData
ItemCount=5}" Margin="0,36,0,97"/>
        <Label x:Name="lblScandItems" Content="Scanned Items:"
HorizontalAlignment="Left"                Margin="10,5,0,0"
VerticalAlignment="Top" FontWeight="Bold" FontSize="14"/>
        <StatusBar x:Name="sBar" Margin="0,371,0,0">
            <TextBlock    x:Name="txtStatus"    TextWrapping="Wrap"
Text="Ready" Height="49" Width="792"/>
        </StatusBar>
    </Grid>
</Window>
```

Here is what the Designer should look like:

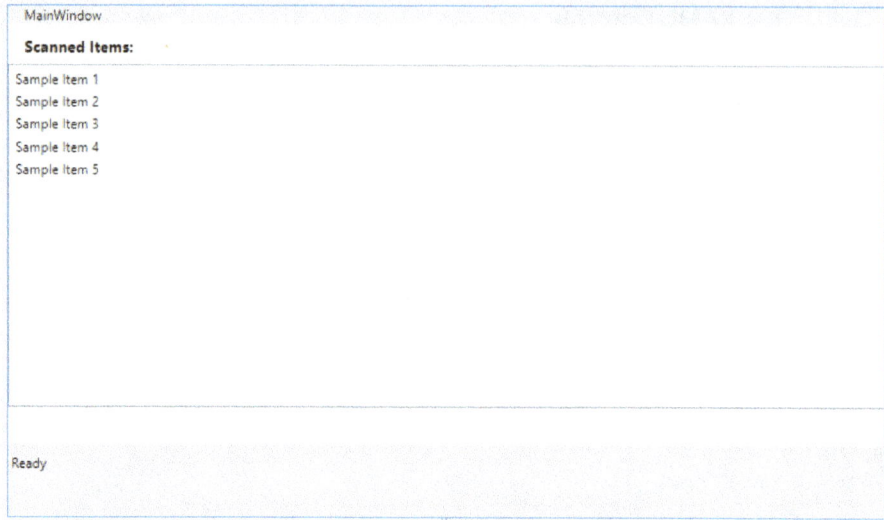

```
MainWindow

  Scanned Items:

Sample Item 1
Sample Item 2
Sample Item 3
Sample Item 4
Sample Item 5




Ready
```

6. Save the MainWindow.xaml file.

6.8.3   Write the code

Now we will put the code behind the interface.

1. Open the MainWindow.xaml.cs file.
2. Add the following using statements so we can access the PointOfService namespace and the namespaces to enumerate the POS devices and decode the barcode scanner input:

```
using Windows.Devices.PointOfService;
using Windows.Devices.Enumeration;
using Windows.Storage.Streams;
```

3. Above the MainWindow() method, add the global properties for a BarcodeScanner, ClaimedBarcodeScanner, and DevcieWatcher.

```
BarcodeScanner scanner = null;
ClaimedBarcodeScanner claimedScanner = null;
DeviceWatcher                 posDevWatcher                 =
DeviceInformation.CreateWatcher(BarcodeScanner.GetDeviceSelecto
r(PosConnectionTypes.All));
```

182

4. After the MainPage() method, create the POSSetup() method and add the following code to create the Device Watcher events:

```
private void POSSetup()
{
    posDevWatcher.Added += PosDevWatcher_Added;
    posDevWatcher.Removed += PosDevWatcher_Removed;
    posDevWatcher.Start();
}
```

Remember when you add the Added and Removed events, after the += hit the Tab key twice to generate the events. The events are listed before the watcher is started. Once the watcher starts, it will continue to monitor for devices being added and removed from the system until either you stop the watcher or the application session is ended.

5. For the PosDevWatcher_Added event handler add the following:

```
private async void PosDevWatcher_Added(DeviceWatcher sender,
DeviceInformation args)
{
    scanner = await BarcodeScanner.FromIdAsync(args.Id);
    if (scanner != null)
    {

        claimedScanner = await scanner.ClaimScannerAsync();
        if (claimedScanner != null)
        {
            scanner.StatusUpdated += Scanner_StatusUpdated;
            claimedScanner.ReleaseDeviceRequested          +=
ClaimedScanner_ReleaseDeviceRequested;
            claimedScanner.DataReceived                    +=
ClaimedScanner_DataReceived;
            claimedScanner.IsDecodeDataEnabled = true;
            await claimedScanner.EnableAsync();

            Application.Current.Dispatcher.Invoke((Action)(()
=> txtStatus.Text = "Barcode Scanner Claimed: " + args.Id));

        }
        else
        {
            Application.Current.Dispatcher.Invoke((Action)(()
=> txtStatus.Text = "Barcode Scanner Not Claimed"));
```

183

```
        }

    }
    else
    {
        Application.Current.Dispatcher.Invoke((Action)(()        =>
txtStatus.Text = "Barcode Scanner Not Found"));
    }
}
```

The Device Watcher added event checks for a barcode scanner being connected and then performs all the same steps to enumerate the device as before.

6. For the PosDevWatcher_Remove event handler add the following:

```
private    void    PosDevWatcher_Removed(DeviceWatcher    sender,
DeviceInformationUpdate args)
{
    claimedScanner.DataReceived -= ClaimedScanner_DataReceived;
    claimedScanner.ReleaseDeviceRequested               -=
ClaimedScanner_ReleaseDeviceRequested;
    scanner.StatusUpdated -= Scanner_StatusUpdated;
    claimedScanner.Dispose();
    scanner.Dispose();
    Application.Current.Dispatcher.Invoke((Action)(()        =>
txtStatus.Text = "Barcode Scanner Removed"));
}
```

The handler actually destroys the claimedScaner and scanner objects by calling the Dispose() method. The Dispose() method could have been used in the previous exercises, but a newly attached device would require a manual operation to connect to the device. Device Watcher will automatically connect to the newly attached barcode scanner.

8. Add the code for the ClaimedScanner_DataReceived event handler method:

```
private  void  ClaimedScanner_DataReceived(ClaimedBarcodeScanner
sender, BarcodeScannerDataReceivedEventArgs args)
{
    //Read the data from the buffer and convert to a string.
    var                  scanDataLabelReader                  =
DataReader.FromBuffer(args.Report.ScanDataLabel);
```

```
    string                    stringDataLabel              =
scanDataLabelReader.ReadString(args.Report.ScanDataLabel.Length
);

    //Since this event is in a different thread, we need to sync
to the UI thread
    Dispatcher.BeginInvoke((Action)(()                        =>
lstItems.Items.Add(stringDataLabel)));
    Dispatcher.BeginInvoke((Action)(()    =>   txtStatus.Text   =
"Ready"));
}
```

9. Add the code for the ClaimedScanner_ReleaseDeviceRequested event handler method:

```
private   void   ClaimedScanner_ReleaseDeviceRequested(object?
sender, ClaimedBarcodeScanner e)
{
    e.RetainDevice();

    Dispatcher.BeginInvoke((Action)(()    =>   txtStatus.Text   =
"Event ReleaseDeviceRequested received. Retaining the barcode
scanner."));
}
```

10. Add the code for the Scanner_StatusUpdated event handler method:

```
private   void   Scanner_StatusUpdated(BarcodeScanner   sender,
BarcodeScannerStatusUpdatedEventArgs args)
{
    uint bcEXStatus = args.ExtendedStatus;
    BarcodeScannerStatus bcStatus = args.Status;
    //Since this event is in a different thread, we need to sync
to the UI thread
    Dispatcher.BeginInvoke((Action)(()    =>   txtStatus.Text    =
"Barcode  status  update: "  +  bcStatus.ToString()  +  " "  +
bcEXStatus.ToString()));
}
```

11. Add POSSetup() to the MainWindow() method:

```
public MainWindow()
```

```
{
    InitializeComponent();
    POSsetup();
}
```

12. Save the project.

6.8.4   Test the Application

Now, we are ready to test the application.

1. In Visual Studio, build the application.
2. If there are any errors, correct them, and rebuild the application.
3. Plug the USB barcode scanner into your development machine, and run the debugger locally. The application will start up. Once the watcher starts, it will connect to the barcode scanner automatically.

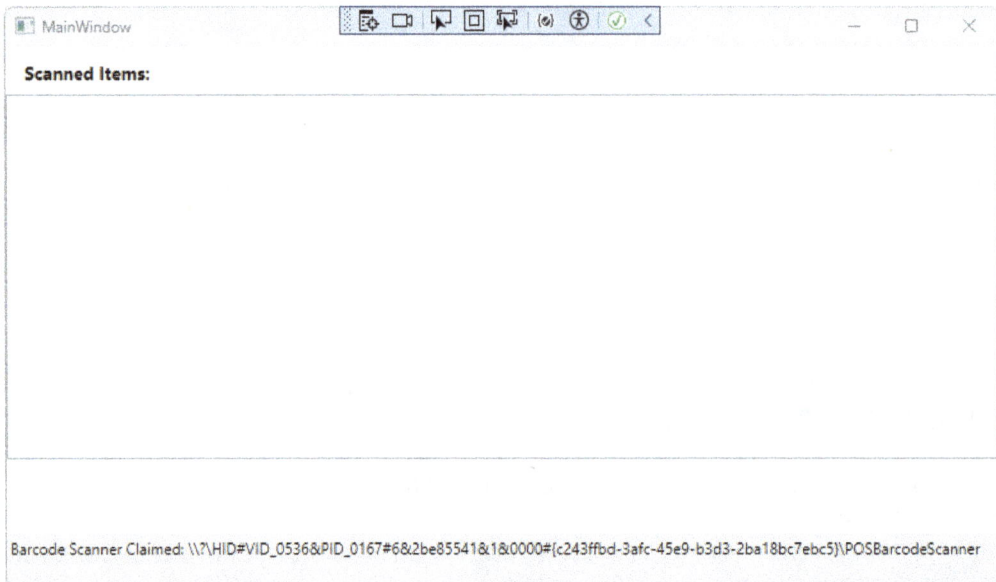

4. Scan a few barcodes.
5. Unplug the USB barcode scanner. The barcode scanner will be removed.

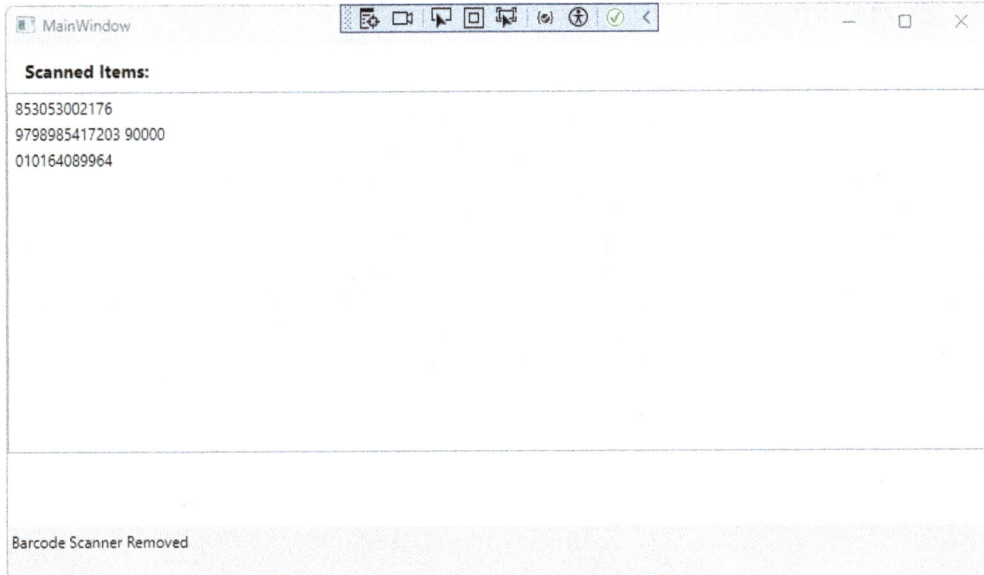

6. Plug the scanner back in. Device Watcher should automatically claim the device.

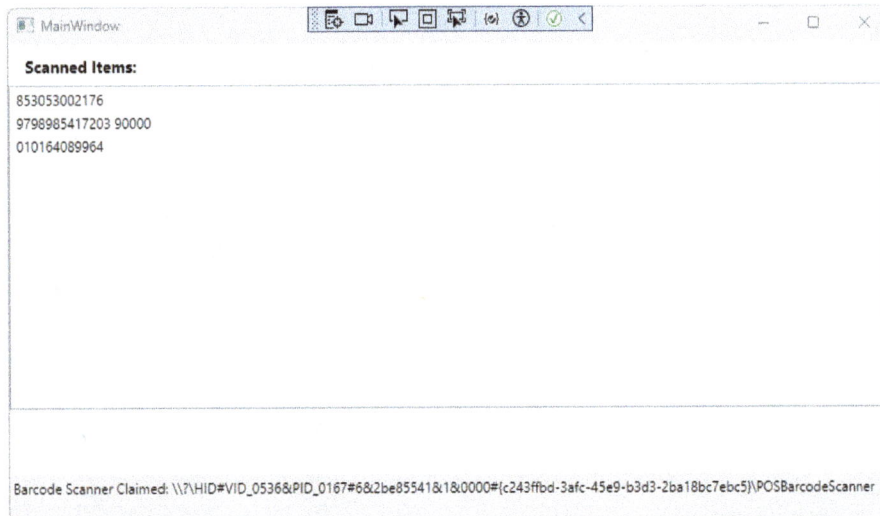

7. Scan a barcode.
8. Stop debugging when finished.

6.9 *Summary*

The POS WinRT namespace allows one to write .NET core applications that can access POS devices. The four enumeration methods provide a standard way to connect to Windows devices. In reality, there are only two enumeration methods that should be considered. The Device Picker was designed for UWP applications, and writing UWP applications is not recommended. The DefaultAsync() method is the most basic method, but should only be used in cases where there is one type of device attached to a system. The last two enumeration methods, Snapshot and Device Watcher, are best for .NET applications. The Snapshot method allows you to create a custom device selection interface, but it takes time to populate the list. Device Watcher is faster executing and matches the DeviceAdded and DeviceRemoved events of POS for .NET. The whole chapter used the barcode scanner to demonstrate all 4 enumeration methods. The next chapter will explore other POS devices.

7 Creating POS WinRT Applications for Different POS Devices

The last chapter covered all four enumeration methods using the barcode scanner. Technically, only two enumeration methods are best for .NET applications. Similar to chapter 3, this chapter will create .NET POS WinRT applications for different POS devices to explore the different WinRT classes and methods. The chapter will cover the following:

- Demonstrate device enumeration for other POS devices.
- Write WPF .NET applications for MSR, Line/Pole Display, Receipt Printer, and cash drawer.
- Create a utility to view all the POS WinRT available devices in a system.

7.1 Exercise 7.1: Magnetic Stripe Reader (MSR) using Device Watcher

For POS WinRT, MSR data is automatically decoded just like the barcode scanner. In addition, POS WinRT supports 3 types of cards:

- Bank Cards
- Driver's licenses or other Vehicle ID cards formatted according to the American Association of Motor Vehicles Administrators (AAMVA) card specification
- Vendor-Specific Cards

In this exercise, we will create an MSR application that uses Device Watcher to get the USB MSR device. The application will support swiping a bank card.

7.1.1 MSR Setup
Hopefully, you have already checked the Barcode scanner in the last chapter. The following steps repeat that setup for a USB MSR device.

189

1. Plug the USB MSR into your system.
2. Open Device Manager and make sure the MSR appears in the list. If the MSR is configured as a keyboard wedge, you will have to follow your manufacturer's instructions to see if it can be switched to a USB /HID device.

7.1.2   Create the Visual Studio Project
The first step is to create the project and perform the basic setup tasks.

10. Open Visual Studio.
11. Create a new project.
12. Search for and select "WPF Application for C#" and click Next.

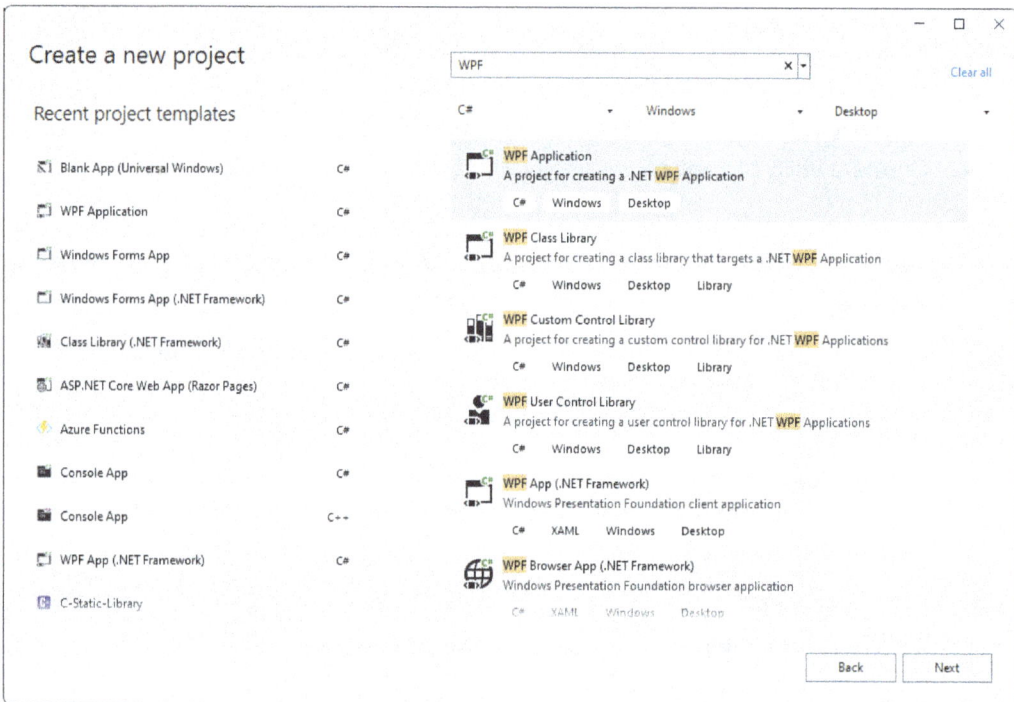

13. Name the project EX71_MSR_DeviceWatcher and click Next.
14. Keep the .NET version at ".NET 8.0 (Long Term Support)" and click Create.

190

15. The next and most important step: from the menu, select Project->
    EX71_MSR_DeviceWatcher properties.
16. Under General, change the following:

> Target OS Version: 10.0.26100.0
> Supported OS Version: 10.0.19041.0

The reason the OS version is important is that the minimum OS version of 10.0.19041.0
is the version that supports the POS WinRT namespaces. If you use any OS version
below the minimum, you will not be able to get access to the POS WinRT namespaces.

17. Save the project when finished.
18. Close the EX71_MST_DeviceWatcher properties page.

7.1.3   Set Up the XAML Controls
Now we will set up the UI.

7. In Solution Explorer, open the MainWindow.xaml.
8. Add a Label in the top, left corner.
    a. Name: lblMSRItems
    b. Font size 14 Bold
    c. Text: Scanned Items
9. Add a ListBox under the label and resize the LisBox to fill 3/4 of the page.
    a. Name: lstItems
10. Add a StatusBar at the bottom of the page.
    a. Name: sBAR
11. Add a TextBlock control to the status strip.
    a. Name: txtStatus

Here is the XAML code:

```
<Window x:Class="EX71_MSR_DeviceWatcher.MainWindow"

xmlns="http://schemas.microsoft.com/winfx/2006/xaml/presentatio
n"
        xmlns:x="http://schemas.microsoft.com/winfx/2006/xaml"

xmlns:d="http://schemas.microsoft.com/expression/blend/2008"
        xmlns:mc="http://schemas.openxmlformats.org/markup-
compatibility/2006"
        xmlns:local="clr-namespace:EX71_MSR_DeviceWatcher"
        mc:Ignorable="d"
        Title="MainWindow" Height="450" Width="800">
```

191

```xml
    <Grid>
        <Label   x:Name="lblMSRResults"   Content="MSR   Results:"
HorizontalAlignment="Left"                       Margin="10,27,0,0"
VerticalAlignment="Top" FontSize="14" FontWeight="Bold"/>
        <ListBox                             x:Name="lstMsrItems"
d:ItemsSource="{d:SampleData ItemCount=5}" Margin="0,61,0,145"/>
        <StatusBar x:Name="sBar" Margin="0,371,0,0">
            <TextBlock   x:Name="txtStatus"   TextWrapping="Wrap"
Text="Ready" Height="49" Width="792"/>
        </StatusBar>
    </Grid>
</Window>
```

The Designer should show the following:

12. Save the MainWindow.xaml file.

7.1.4 Write the code
Now we will create the code behind the interface.

15. Open the MainWindow.xaml.cs file.
16. Add the following using statements so we can access the PointOfService namespace and enumeration namespace:

192

```
using Windows.Devices.PointOfService;
using Windows.Devices.Enumeration;
using Windows.Storage.Streams;
```

17. Above the MainWindow() method, add the global instances for a MagneticStripeReader, ClaimedMagneticStripeReader, and DeviceWatcher classes. These will be used in the methods of MainWindow class.

```
MagneticStripeReader myMSR = null;
ClaimedMagneticStripeReader claimedMSR = null;
DeviceWatcher                      posDevWatcher                      =
DeviceInformation.CreateWatcher(MagneticStripeReader.GetDeviceS
elector(PosConnectionTypes.All));
```

The same discussion about the barcode reader classes from the last chapter applies to the MagneticStripeReader and ClaimedMagneticStripeReader. The MagneticStripeReader class is used to make the connection, and the ClaimedMagneticStripeReader class contains all the data event handling.

18. After the MainWindow() method, create the POSSetup() method and add the following code to create the Device Watcher:

```
private void POSSetup()
{
    posDevWatcher.Added += PosDevWatcher_Added;
    posDevWatcher.Removed += PosDevWatcher_Removed;
    posDevWatcher.Start();
}
```

The device watcher is set up to search for MSR devices only, on any POS Connection Type.

19. For the MsrDevWatcher_Added event handler, add the following:

```
private async void PosDevWatcher_Added(DeviceWatcher sender,
DeviceInformation args)
{
    myMSR = await MagneticStripeReader.FromIdAsync(args.Id);
    if (myMSR != null)
    {
        claimedMSR = await myMSR.ClaimReaderAsync();
        if (claimedMSR != null)
        {
```

```
                myMSR.StatusUpdated += MyMSR_StatusUpdated;
                claimedMSR.ReleaseDeviceRequested            +=
ClaimedMSR_ReleaseDeviceRequested;
                claimedMSR.BankCardDataReceived              +=
ClaimedMSR_BankCardDataReceived;
                claimedMSR.IsDecodeDataEnabled = true;
                await claimedMSR.EnableAsync();
                Application.Current.Dispatcher.Invoke((Action)(()
=> txtStatus.Text = "MSR Claimed: " + args.Id));


            }
            else
            {
                Application.Current.Dispatcher.Invoke((Action)(()
=> txtStatus.Text = "MSR Not Claimed"));
            }
        }
        else
        {
            Application.Current.Dispatcher.Invoke((Action)(()      =>
txtStatus.Text = "MSR Not Found"));
        }
    }
}
```

With the barcode scanner, we simply added the device to a list box control from which the user selects the device to connect. The MsrDevWatcher_Added event above is acting to find the first available MSR and completing the process to claim and set up the data received and other events. The only problem with this is if a second MSR is plugged in, it will break the first MSR. In this case, the MsrDevWatcher_Added addresses this situation.

20. For the MsrDevWatcher_Remove event handler, add the following:

```
private    void    PosDevWatcher_Removed(DeviceWatcher    sender,
DeviceInformationUpdate args)
{
    claimedMSR.BankCardDataReceived                              -=
ClaimedMSR_BankCardDataReceived;
    claimedMSR.ReleaseDeviceRequested                           -=
ClaimedMSR_ReleaseDeviceRequested;
    myMSR.StatusUpdated -= MyMSR_StatusUpdated;
    claimedMSR.Dispose();
    myMSR.Dispose();
```

194

```
    Application.Current.Dispatcher.Invoke((Action)(()          =>
txtStatus.Text = "Msr Removed"));
}
```

The MsrDevWatcher_Remove event removes the events and disposes of the claim on the MSR. Now, let's finish the events created from the MsrDevWatcher_Added event.

21. Enter the following code into the MyMSR_StatusUpdate event handler, which is similar to the barcode scanner:

```
private void MyMSR_StatusUpdated(MagneticStripeReader sender,
MagneticStripeReaderStatusUpdatedEventArgs args)
{
    uint msrEXStatus = args.ExtendedStatus;
    MagneticStripeReaderStatus msrStatus = args.Status;
    Dispatcher.Invoke((Action)(()  =>  txtStatus.Text  =  "MSR
status update: " + msrStatus.ToString()));
}
```

22. Add the following code to the ClaimedMSR_ReleaseDeviceRequestedevent handler, which is similar to the barcode scanner:

```
private void ClaimedMSR_ReleaseDeviceRequested(object? sender,
ClaimedMagneticStripeReader e)
{
    Dispatcher.BeginInvoke((Action)(() => txtStatus.Text = "MSR
being released"));
}
```

23. Add the following code to the ClaimedMSR_BankCardDataReceivedevent handler:

```
private                                                      void
ClaimedMSR_BankCardDataReceived(ClaimedMagneticStripeReader
sender, MagneticStripeReaderBankCardDataReceivedEventArgs args)
{
    Dispatcher.BeginInvoke((Action)(()                       =>
lstMsrItems.Items.Add(args.AccountNumber)));
    //Dispatcher.BeginInvoke((Action)(()                     =>
lstMsrItems.Items.Add(args.ExpirationDate)));
    //Dispatcher.BeginInvoke((Action)(()                     =>
lstMsrItems.Items.Add(args.FirstName)));
```

195

```
    //Dispatcher.BeginInvoke((Action)(()                    =>
lstMsrItems.Items.Add(args.MiddleInitial)));
    //Dispatcher.BeginInvoke((Action)(()                    =>
lstMsrItems.Items.Add(args.Surname)));
    //Dispatcher.BeginInvoke((Action)(()                    =>
lstMsrItems.Items.Add(args.ServiceCode)));
    //Dispatcher.BeginInvoke((Action)(()                    =>
lstMsrItems.Items.Add(args.Suffix)));
    //Dispatcher.BeginInvoke((Action)(()                    =>
lstMsrItems.Items.Add(args.Title)));
}
```

POS for .NET required an ASCIIEncoding conversion to string to get the data from a specific track. Forgetting the Dispatcher requirement to interact with the UI, the MagneticStripRederBankCardDataReceivedEventArgs simply provides the data in one line of code, which is the account number in this case. You can get all sorts of data from the bank card by simply uncommenting the different lines of code for First Name, Surname, Expiration Data, etc.

24. In the MainWindow() method, add the call to the POSSetup() method:

```
public MainWindow()
{
    InitializeComponent();
    POSSetup();
}
```

25. Remove any unused "using" declarations.
26. Save the whole project.

7.1.5   Test the Application
Now, we are ready to test the application.

1. In Visual Studio, build the application.
2. If there are any errors, correct them, and rebuild the application.
3. Plug the USB MSR into your development machine, and run the debugger. The application will start and claim the MSR.

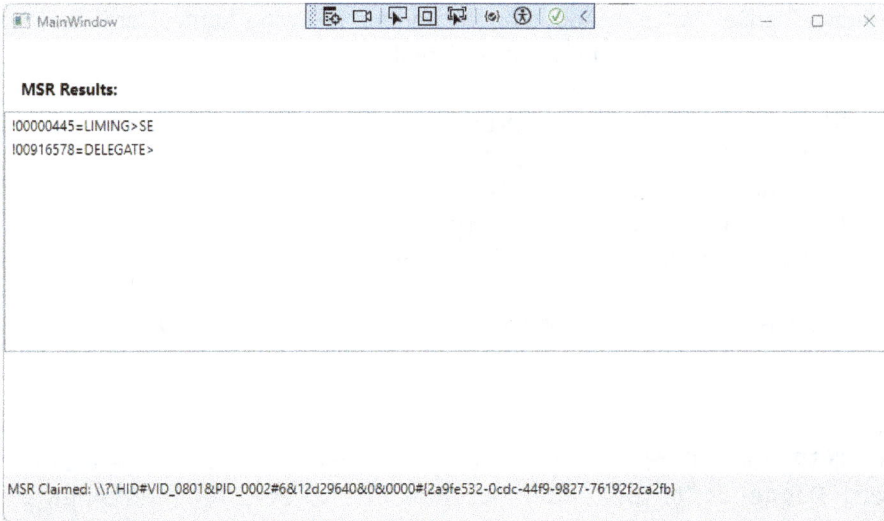

4. Swipe a few cards.
5. Unplug the MSR and plug it back in.

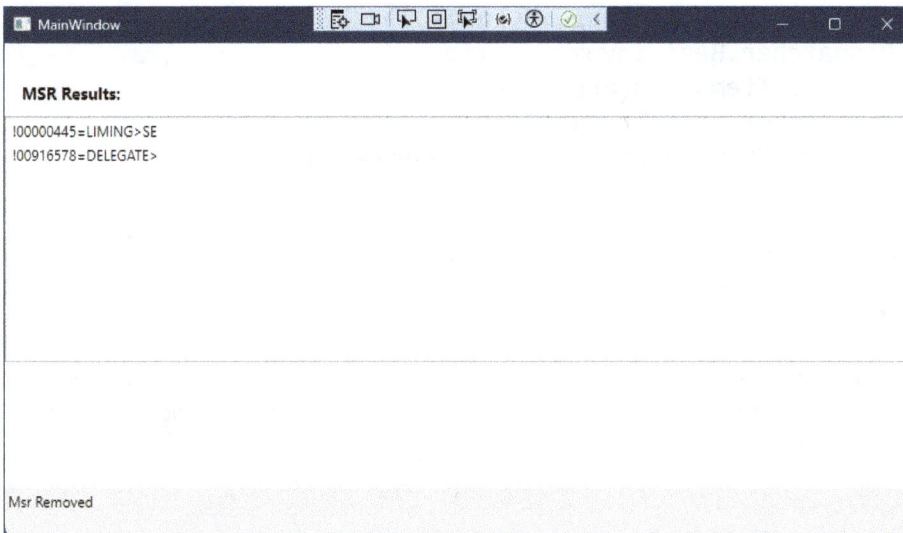

6. Stop debugging when finished.

7.1.6   Driver License Test

Let's add support to get swipe data from driver licenses.

1. In the MsrDevWatcher_Added event, add the following AamvaCardDataReceived event after the line for the BankCardDataReceived event:

```
claimedMSR.BankCardDataReceived                                    +=
ClaimedMSR_BankCardDataReceived;
claimedMSR.AamvaCardDataReceived                                   +=
ClaimedMSR_AamvaCardDataReceived;
claimedMSR.IsDecodeDataEnabled = true;
```

2. Enter the following code for the ClaimedMSR_AamvaCardDataReceived event handler:

```
private                                                             void
ClaimedMSR_AamvaCardDataReceived(ClaimedMagneticStripeReader
sender, MagneticStripeReaderAamvaCardDataReceivedEventArgs args)
{
    Dispatcher.BeginInvoke((Action)(()                             =>
lstMsrItems.Items.Add(args.FirstName)));
    Dispatcher.BeginInvoke((Action)(()                             =>
lstMsrItems.Items.Add(args.Surname)));
    Dispatcher.BeginInvoke((Action)(()                             =>
lstMsrItems.Items.Add(args.EyeColor)));
    Dispatcher.BeginInvoke((Action)(()                             =>
lstMsrItems.Items.Add(args.LicenseNumber)));
}
```

Like the bank card, the driver's license information is automatically decoded. There is more data than what the code above will display; but if you are interested, you can make any changes to test this out.

3. Run the debugger and swipe a driver's license card. The driver's license data will appear in the list box. Notice that the Bank Card event didn't trigger as the classes know the differences between the card types.

7.2 Exercise 7.2 Line/Pole Display Using Device Watcher

A Line or Pole Display is the simplest POS device, since all it does is display text. This exercise will recreate a .NET WPF application that sends a message to the Pole Display. POS WinRT's support for the Line or Pole Display is through a line display's OPOS driver. Since an OPOS driver is needed, the real work is in the setup of the OPOS drivers.

7.2.1 OPOS Driver Setup

The EPSON DM-D110 line display will be used for this exercise. Unlike the Chapter 3 Pole Display applications, the Pole Display will be connected via the USB port instead of the serial port. The USB port will be treated as a COM port. The nice thing about using the USB port is that power to the DM-D110 comes from the USB and the extra power supply is not needed.

1. If you install the Epson OPOS ADK for .NET v1.14.30E, the whole ADK and its configurations need to be removed. Run the setup utility and delete any and all setup devices.
2. Run POSDM listdevices or SOManager to make sure all the Epson devices are removed.
3. Uninstall, the Epson OPOS ADK for .NET v1.14.30E
4. For the DM-D110, A driver and a different OPOS ADK are needed for POS WinRT OPOS bridge driver. Open a browser and search for "Epson DM-D110".
5. At the bottom of the EPSON DM-D110 product page, set the search for Windows 10 32-bit:

   - Epson OPOS ADK v3.00ER23, which is a different OPOS driver.
   - DM-D Virtual COM Port Driver v2.02 (Search in Windows 10 32 since it is not available for 64-bit, and scroll down the support page for this one)

6. Plug the DM-D110 into your target system.
7. Extract the DM-D Virtual COM Port Driver.
8. Open Device Manager, and you will see that a driver is missing.
9. Update the missing driver by browsing to the location of the extracted DM-D Virtual COM Port Driver. This will be the USB bus driver.

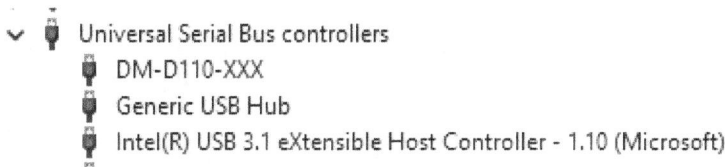

10. Once the first driver has been installed, a second driver will show up as missing.
11. Update the missing driver by browsing to the location of the extracted DM-D Virtual COM Port Driver. You should see the DM-D110-XXX appear under ports. Make a note of the COM port number as we will set up the OPOS driver to use that COM port.

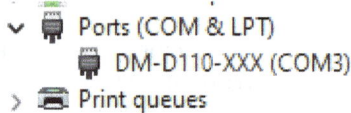

> ∨ 🖶 Ports (COM & LPT)
>   🖶 DM-D110-XXX (COM3)
> › 🖨 Print queues

12. Install the EPSON OPOS ADK.
13. After installation, run the new SetupPOS utility.
14. Expand the device tree on the left.
15. Right-click on "LineDisplay", and select "Add New Device…".
16. The first dialog will ask you for the Logic Device Name, enter myLineDisplay, and hit Next.
17. After a few moments, a popup dialog will ask if this is standalone or connected via another device. Keep the default standalone and click OK.
18. The next screen will appear asking for you to set the COM port. Use the drop-down to select the COM port that was noted from the Device Manager.

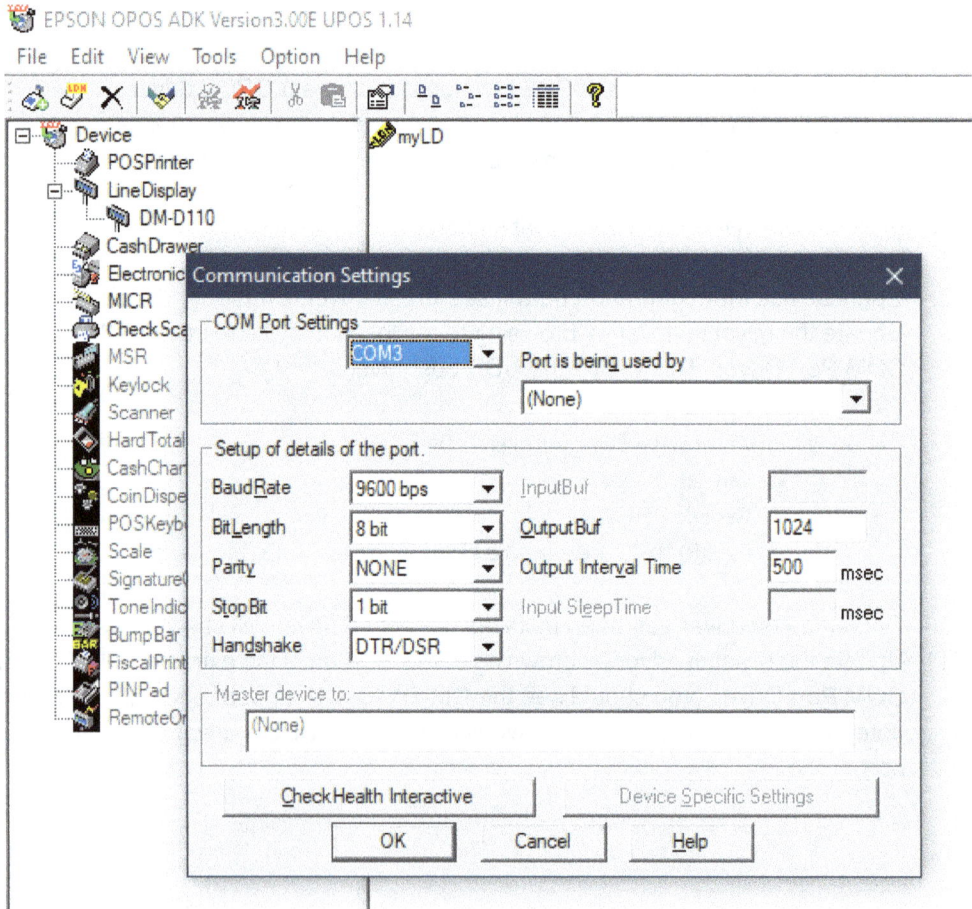

19. Click Finish.
20. Restart the computer.

7.2.2 Create the Visual Studio Project
The first step is to create the project and perform the basic setup tasks.

1. Open Visual Studio.
2. Create a new project.
3. Search for and select "WPF Application for C#" and click Next.
4. Name the project EX72_PoleDisplay_DeviceWatcher and click Next.
5. Keep the .NET version at ".NET 8.0 (Long Term Support)" and click Create.
6. The next and most important step: from the menu select Project-> EX72_PoleDisplay_DeviceWatcher properties.
7. Under General, change the following:

> Target OS Version: 10.0.26100.0
> Supported OS Version: 10.0.19041.0

The reason the OS version is important is that the minimum OS version of 10.0.19041.0 is the version that supports the POS WinRT namespaces. If you use any OS version below the minimum, you will not be able to get access to the POS WinRT namespaces.

8. Save the project when finished.
9. Close the EX72_ PoleDisplay _DeviceWatcher properties page.

7.2.3 Set Up the XAML Controls
Now we will set up the UI.

1. In Solution Explorer, open the MainWindow.xaml.
2. Add a TextBox control.
   a. Name: txtInput
3. Add a Button.
   a. Name: btnSend
   b. Content: Send to pole Display
4. Add the Status Strip, resize to fit the bottom of the page
   a. Name: sBar
5. Add a TextBlock control to the status strip
   b. Name: txtStatus
   c. Font 10 px

Here is the XAML code:

```xml
<Window x:Class="EX72_PoleDisplay_DefaulAsync.MainWindow"

xmlns="http://schemas.microsoft.com/winfx/2006/xaml/presentatio
n"
        xmlns:x="http://schemas.microsoft.com/winfx/2006/xaml"

xmlns:d="http://schemas.microsoft.com/expression/blend/2008"
        xmlns:mc="http://schemas.openxmlformats.org/markup-
compatibility/2006"
        xmlns:local="clr-
namespace:EX72_PoleDisplay_DefaulAsync"
        mc:Ignorable="d"
        Title="MainWindow" Height="450" Width="800">
    <Grid>
        <TextBox   x:Name="txtInput"   HorizontalAlignment="Left"
Margin="92,88,0,0" TextWrapping="Wrap" Text="Enter a  message
here" VerticalAlignment="Top" Width="633"/>
        <Button x:Name="btnSend" Content="Send to Pole Display"
HorizontalAlignment="Left"             Margin="571,178,0,0"
VerticalAlignment="Top"          Width="134"          Height="54"
Click="btnSend_Click"/>
        <StatusBar x:Name="sBar" Margin="0,380,0,0">
            <TextBlock   x:Name="txtStatus"   TextWrapping="Wrap"
Text="Ready" Width="778" Height="32"/>
        </StatusBar>

    </Grid>
</Window>
```

The user will simply enter some text in the upper textbox and push the button to send the data to the Line Display. The Designer should show the following:

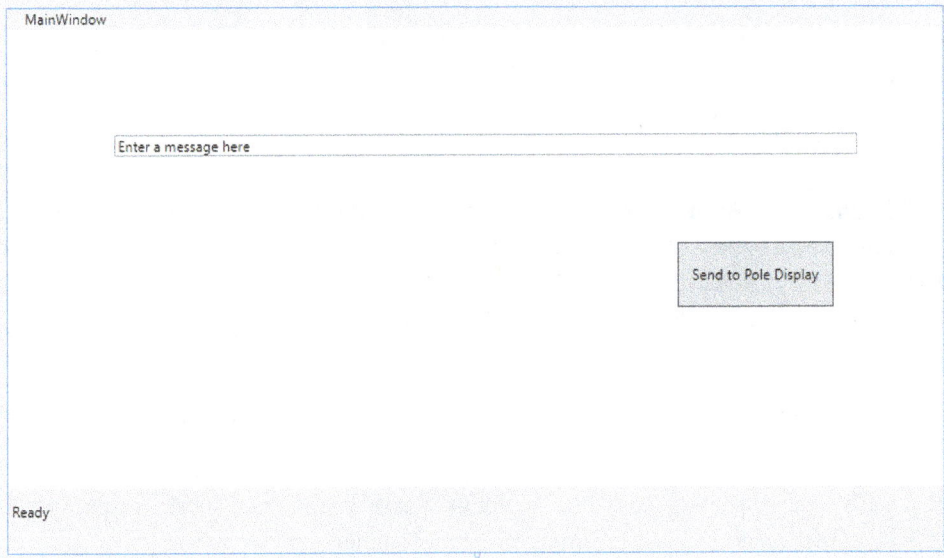

6. Save the MainWindow.xaml file.

7.2.4   Write the code
Now we will create the code behind the interface.

1. Open the MainWindow.xaml.cs file.
2. Add the following using statements so we can access the PointOfService namespace:

```
using Windows.Devices.PointOfService;
using Windows.Devices.Enumeration;
```

3. Above the MainWindow() method add the global instances for LineDisplay and ClaimedLineDisplay classes. Also, add the device watcher for the line display. These will be used in the methods of the MainWindow class.

```
LineDisplay myLineDisplay = null;
ClaimedLineDisplay myClaimedLineDisplay= null;
DeviceWatcher                lineDisplayDevWatcher               =
DeviceInformation.CreateWatcher(LineDisplay.GetDeviceSelector(P
osConnectionTypes.All));
```

LineDisplay class helps with making the connection. ClaimedLineDisplay handles any events or actions with the POS device.

203

4. After the MainWindow() method, create the POSSetup() method and add the following code to connect and claim the Line display.

```
public void POSSetup()
{
    lineDisplayDevWatcher.Added += LineDisplayDevWatcher_Added;
    lineDisplayDevWatcher.Removed                            +=
LineDisplayDevWatcher_Removed;
    lineDisplayDevWatcher.Start();
}
```

POSSetup adds the device watcher added and removed events and then starts the device watcher for the line display.

5. Let's fill in the device watcher events. First, add code to the LineDisplayDevWatcher_Removed event handler:

```
private void LineDisplayDevWatcher_Removed(DeviceWatcher sender,
DeviceInformationUpdate args)
{
    myClaimedLineDisplay.Dispose();
    myLineDisplay.Dispose();
}
```

6. Add the code for the LineDisplayDevWatcher_Added event handler.

```
private async void LineDisplayDevWatcher_Added(DeviceWatcher
sender, DeviceInformation args)
{
    myLineDisplay = await LineDisplay.FromIdAsync(args.Id);
    if (myLineDisplay != null) {

        myClaimedLineDisplay = await myLineDisplay.ClaimAsync();
        if (myClaimedLineDisplay != null) {

            myClaimedLineDisplay.ReleaseDeviceRequested        +=
MyClaimedLineDisplay_ReleaseDeviceRequested;
            Dispatcher.Invoke((Action)(()  =>  txtStatus.Text  =
txtStatus.Text = "Line Display is claimed and enabled"));
        }
        else
        {
```

```
        Dispatcher.Invoke((Action)(() => txtStatus.Text =
"Line Display is NOT claimed"));
        }
    }
    else
    {
        Dispatcher.Invoke((Action)(() => txtStatus.Text = "Line
Display is NOT found"));
    }
}
```

Like previous device watcher solutions, the code waits for a line display. If found, the code claims the display. There is no Enable property so we just end with claim. Any errors will be sent to the status bar.

7. Since it was created by the device watcher added event handler, fill in the code for the MyClaimedLineDisplay_ReleaseDeviceRequested event hanlder:

```
private                                                    void
MyClaimedLineDisplay_ReleaseDeviceRequested(ClaimedLineDisplay
sender, object args)
{
    sender.RetainDevice();
}
```

8. Go back to the MainWindow.xaml, and in the Designer, double-click on the button to generate a button event handler in the ManagePage.xaml.cs file.
9. Add the following code to the btnSend_Click:

```
private async void btnSend_Click(object sender, RoutedEventArgs
e)
{
    bool          clearSuccess          =          await
myClaimedLineDisplay.DefaultWindow?.TryClearTextAsync();
    if (clearSuccess)
    {
        txtStatus.Text = "Displayed cleared sucessfully";
    }
    else
    {
        txtStatus.Text = "Display clear failed";
    }
```

```
    LineDisplayTextAttribute                attribute              =
LineDisplayTextAttribute.Normal;
    bool            displaySuccess            =            await
myClaimedLineDisplay.DefaultWindow.TryDisplayTextAsync(txtInput
.Text.ToString(), attribute);
    if (displaySuccess)
    {
        txtStatus.Text += " Data sent sucessfull to the display";
    }
    else {
        txtStatus.Text += " Attempt to display failed";
    }
}
```

When the button is pressed, the line display will be cleared, and the code checks the return status to see if anything failed. Next, the display attribute is set to communicate to the display how to display the text. There are 4 options: Blink, Normal, Reverse, and Reverse Blink. Both of the reverse options are for when the Line Display is hung upside down. Finally, we send the text in the textbox to the display. A final check is made to see if the data was sent successfully. If you look at the call to myClaimedDisplay.DefaultWindow.TryDisplayTextAsync and check the online documentation, you will see that DefaultWindow is a property of the ClaimedLineDisplay class. TryDisplayTextASync is a method under the LineDisplayWindow Class. It is not clear on the hierarchy of the class design how DefaultWindow is linked to the LineDisplayWindow Class based on the documentation. Regardless of how the classes are designed, you can see that the LineDisplayWindow class has other methods to send Bitmaps, scroll text, etc.

10. In the main method, add the call to the POSSetup() method:

```
public MainWindow()
{
    InitializeComponent();
    POSSetup();
}
```

11. Remove any unused "using" declarations.
12. Save the whole project.

7.2.5   Test the Application
Now, we are ready to test the application.

1.   In Visual Studio, build the application.

2. If there are any errors, correct them, and rebuild the application.
3. Plug the Line Display into your development machine, and run the debugger locally.
4. Once the application starts, the Line Display should be claimed.

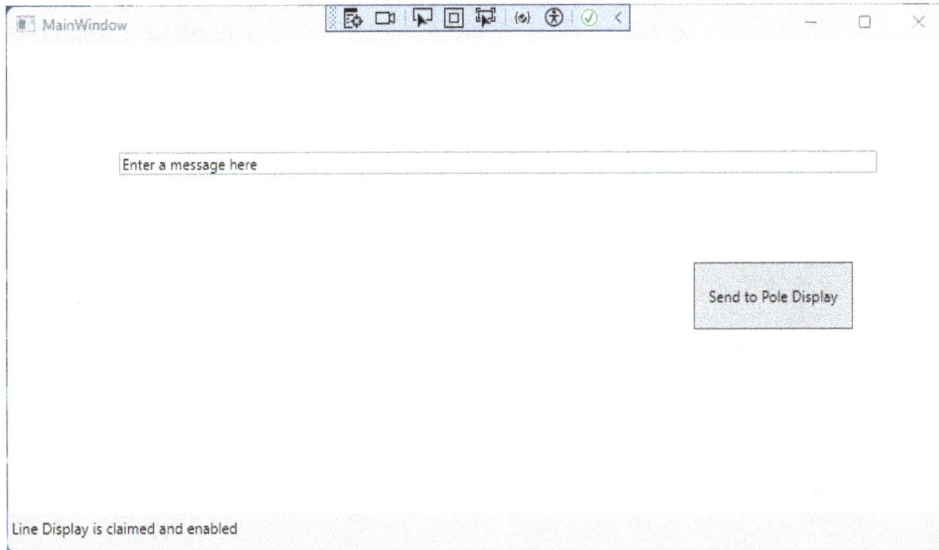

5. Enter some text in the text box and press the button. The line display device will show the text entered.
6. Stop the debugger when finished.

7.3 Exercise 7.3: POS Receipt Printer and Cash Drawer using Device Watcher

The Epson TM-T88V POS printer and the MMF cash drawer will be used in this exercise. For this exercise, we will make the connection over Ethernet. The MMF cash drawer will be connected to the DK port of the TM-T88V. The TM-T88V has the UB-E04 Ethernet interface module.

If you want to use the USB, the USB connection requires a virtual COM Port driver and the EPSON OPOS ADK driver, not the EPSON OPOS ADK for .NET driver. The ADK setup utility has to configure both POS devices. IF you didn't do the USB Receipt Printer application in chapter 3, the 88V has the USB port hidden behind a cover.

7.3.1 Printer Setup

If you need to configure the interface for Ethernet, hold the line-feed button down on the printer and power on the printer. Once the first menu appears, release the line-feed button.

Then hold the line-feed button down again, until the menu prints out. Tap the line-feed button 3 times for "Customize Value Settings", and press and hold the line-feed button until the next menu is printed out. Interface selection should be number 10. Tap the line-feed button 10 times, and then press and hold the line-feed button until the Interface selection menu is printed out. The setting needs to be set to Auto. Tap the line-feed button as many times as needed to reach Auto, and then press and hold the line-feed button until the printer prints confirmation of the selection. Power off the printer. When you power on the printer, the printer will print out its TCP/IP address.

7.3.2   Create the Visual Studio Project
The first step is to create the project and perform the basic setup tasks.

1. Open Visual Studio.
2. Create a new project.
3. Search for and select "WPF Application for C#" and click Next.
4. Name the project EX73_PrinterCD_DeviceWatcher and click Next.
5. Keep the .NET version at ".NET 8.0 (Long Term Support)" and click Create.
6. The next and most important step: from the menu select Project-> EX73_PrinterCD_DeviceWatcher properties.
7. Under General, change the following:

> Target OS Version: 10.0.26100.0
> Supported OS Version: 10.0.19041.0

The reason the OS version is important is that the minimum OS version of 10.0.19041.0 is the version that supports the POS WinRT namespaces. If you use any OS version below the minimum, you will not be able to get access to the POS WinRT namespaces.

8. Save the project when finished.
9. Close the EX73_PrinterCD_DeviceWatcher properties page.

7.3.3   Set Up the XAML Controls
Now we will set up the UI.

1. In Solution Explorer, open the MainWindow.xaml.
2. Add a TextBox control.
   a. Name: txtInput
3. Add a Button.
   a. Name: btnPrint
   b. Content: Print Message
4. Add a Button.
   a. Name: btnOpenCD
   b. Content: Open Cash Drawer

5. Add the Status strip, resize to fit the bottom of the page
   a. Name: sBar
6. Add a TextBlock control to the status strip
   a. Name: txtStatus
   b. Text: Ready
   c. Font: 10px

Here is the XAML code:

```xml
<Window x:Class="EX72_PrinterCD_DeviceWatcher.MainWindow"

xmlns="http://schemas.microsoft.com/winfx/2006/xaml/presentatio
n"
        xmlns:x="http://schemas.microsoft.com/winfx/2006/xaml"

xmlns:d="http://schemas.microsoft.com/expression/blend/2008"
        xmlns:mc="http://schemas.openxmlformats.org/markup-
compatibility/2006"
        xmlns:local="clr-
namespace:EX72_PrinterCD_DeviceWatcher"
        mc:Ignorable="d"
        Title="MainWindow" Height="450" Width="800">
    <Grid>
        <TextBox  x:Name="txtInput"  HorizontalAlignment="Left"
Margin="94,87,0,0"  TextWrapping="Wrap"  Text="Enter  a  message
here" VerticalAlignment="Top" Width="562"/>
        <Button    x:Name="btnPrint"    Content="Print    Message"
HorizontalAlignment="Left"              Margin="94,136,0,0"
VerticalAlignment="Top"           Width="153"           Height="71"
Click="btnPrint_Click"/>
        <Button  x:Name="btnOpenCD"  Content="Open  Cash  Drawer"
HorizontalAlignment="Left"              Margin="94,234,0,0"
VerticalAlignment="Top"          Height="71"          Width="153"
Click="btnOpenCD_Click"/>
        <StatusBar x:Name="sBar" Margin="0,367,0,0">
            <TextBlock   x:Name="txtStatus"   TextWrapping="Wrap"
Text="Ready" Width="785" Height="58" FontSize="10"/>
        </StatusBar>
    </Grid>
</Window>
```

The user will simply enter some text in the upper textbox and push the button to send the data to the printer. The open cash drawer button will simply open the cash drawer. The Designer should show the following:

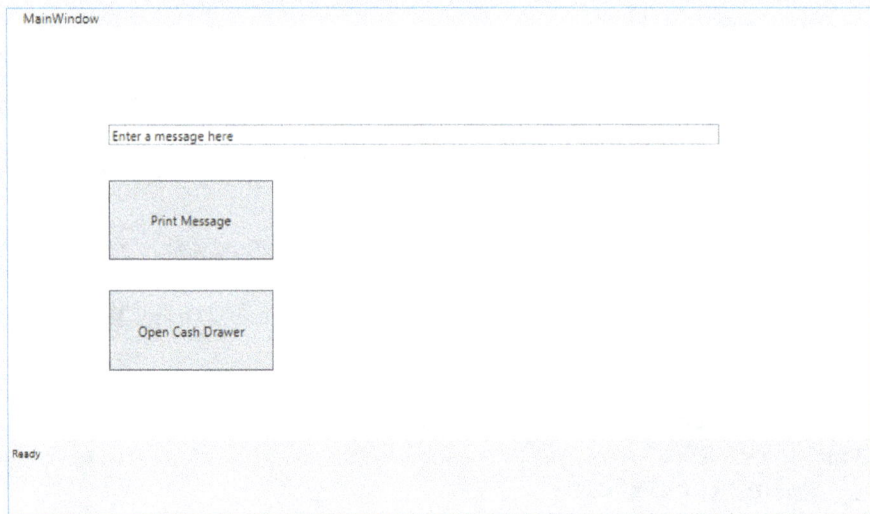

7. Save the MainWindow.xaml file.

7.3.4   Write the code
Now we will create the code behind the interface.

1. Open the MainWindow.xaml.cs file.
2. Add the following using statements, so we can access the PointOfService and Enumeration namespaces:

```
using Windows.Devices.PointOfService;
using Windows.Devices.Enumeration;
```

3. Above the MainWindow() method add the global instances for a PosPrinter, ClaimedPosPrinter, CashDrawer, ClaimedCashDrawer, and 2 Device Watcher classes. These will be used in the methods of the MainWindow class.

```
PosPrinter myPrinter = null;
ClaimedPosPrinter myClaimedPrinter = null;
CashDrawer myCashDrawer = null;
ClaimedCashDrawer myClaimedCD = null;
```

210

```
DeviceWatcher                    printerDevWatcher                =
DeviceInformation.CreateWatcher(PosPrinter.GetDeviceSelector(Po
sConnectionTypes.All));
DeviceWatcher               cashdrawerDevWatcher                 =
DeviceInformation.CreateWatcher(CashDrawer.GetDeviceSelector(Po
sConnectionTypes.All));
```

4. After the MainWindow() method, create the POSSetup() method to add the two device watchers:

```
private void POSSetup()
{
    printerDevWatcher.Added += PrinterDevWatcher_Added;
    printerDevWatcher.Removed += PrinterDevWatcher_Removed;
    printerDevWatcher.Start();

    cashdrawerDevWatcher.Added += CashdrawerDevWatcher_Added;
    cashdrawerDevWatcher.Removed                            +=
CashdrawerDevWatcher_Removed;
}
```

Two device watchers are needed for each device. It would have been nice to have one, but there is no device selector or DeviceClass that filters two POS devices. The add event handlers will perform all the actions to connect to the printer and cash drawer automatically. The printer's device watcher will be started, and once the printer has been claimed, the Cash Drawer device watcher will start. The next step is to set up all the event handlers.

5. Enter the following for the MyClaimedPrinter_ReleaseDeviceRequested event handler:

```
private                                               void
MyClaimedPrinter_ReleaseDeviceRequested(ClaimedPosPrinter
sender, PosPrinterReleaseDeviceRequestedEventArgs args)
{
    Dispatcher.BeginInvoke((Action)(()  =>  txtStatus.Text  =
"Printer being released"));
}
```

6. Enter the following for the PrinterDevWatacher_Added event handler:

```
private async void PrinterDevWatcher_Added(DeviceWatcher sender,
DeviceInformation args)
{
```

211

```
        //Dispatcher.Invoke((Action)(() => txtStatus.Text = "Found
printer: " + args.Name + " | " + args.Id));

    myPrinter = await PosPrinter.FromIdAsync(args.Id);
    if (myPrinter != null)
    {

        myClaimedPrinter = await myPrinter.ClaimPrinterAsync();

        if (myClaimedPrinter != null)
        {
            myClaimedPrinter.ReleaseDeviceRequested            +=
MyClaimedPrinter_ReleaseDeviceRequested;
            if (await myClaimedPrinter.EnableAsync())
            {
                Dispatcher.Invoke((Action)(() => txtStatus.Text
= txtStatus.Text += "Printer is claimed and enabled"));
                cashdrawerDevWatcher.Start();
            }
            else
            {
                Dispatcher.Invoke((Action)(() => txtStatus.Text
= "Printer is NOT claimed and enabled"));
            }
        }
        else
        {
            Dispatcher.Invoke((Action)(()  =>  txtStatus.Text  =
"Claim Printer failed"));
        }
    }
    else
    {
        Dispatcher.Invoke((Action)(()    =>    txtStatus.Text    =
"Printer Not Found"));
    }
}
```

7. Enter the following for the PrinterDevWatacher_Removed event handler:

```
private  void  PrinterDevWatcher_Removed(DeviceWatcher  sender,
DeviceInformationUpdate args)
```

212

```
{
    myClaimedPrinter.ReleaseDeviceRequested                    -=
MyClaimedPrinter_ReleaseDeviceRequested;
    myClaimedPrinter.Dispose();
    myPrinter.Dispose();

    Dispatcher.BeginInvoke((Action)(()    =>    txtStatus.Text    =
"Printer Removed: " + args.Id));
}
```

8. Enter the following for the MyClaimedCD_ReleaseDeviceRequested event handler:

```
private                                                        void
MyClaimedCD_ReleaseDeviceRequested(ClaimedCashDrawer      sender,
object args)
{
    Dispatcher.BeginInvoke((Action)(() => txtStatus.Text = "Cash
Drawer being released"));
}
```

9. Enter the following for the CashDrawerDevWatcher_Added event handler:

```
private  async  void  CashdrawerDevWatcher_Added(DeviceWatcher
sender, DeviceInformation args)
{

    //Dispatcher.Invoke((Action)(()  =>  txtStatus.Text  =  "Found
cash drawer " + args.Name + " | " + args.Id));

    myCashDrawer = await CashDrawer.FromIdAsync(args.Id);

    if (myCashDrawer != null)
    {

        myClaimedCD = await myCashDrawer.ClaimDrawerAsync();
        if (myClaimedCD != null)
        {
            myClaimedCD.ReleaseDeviceRequested              +=
MyClaimedCD_ReleaseDeviceRequested;
            if (await myClaimedCD.EnableAsync())
            {
```

213

```
                Dispatcher.Invoke((Action)(() => txtStatus.Text
= txtStatus.Text += " Cash Drawer is claimed and enabled"));
            }
        else
        {
                Dispatcher.Invoke((Action)(() => txtStatus.Text
= txtStatus.Text += " Cash Drawer is NOT claimed and enabled"));
            }
        }
    }
    else
    {
        Dispatcher.Invoke((Action)(()     =>    txtStatus.Text    =
txtStatus.Text += " Cash Drawer not found"));
    }
}
```

10. Enter the following for the CashDrawerDevWatcher_Removed event handler:

```
private void CashdrawerDevWatcher_Removed(DeviceWatcher sender,
DeviceInformationUpdate args)
{
    myClaimedCD.ReleaseDeviceRequested                          -=
MyClaimedCD_ReleaseDeviceRequested;
    myClaimedCD.Dispose();
    myCashDrawer.Dispose();
    Dispatcher.Invoke((Action)(()    =>   txtStatus.Text   =   "Cash
Drawer Removed"));
}
```

11. Go back to the MainWindow.xaml; and in the Designer, double-click on the buttons to generate the button event handlers in the MainWindow.xaml.cs file.
12. In the MainWindow.xaml.cs file, add an asynchronous method to the btnPrint_Click and btnOpenCashDrawer_Click declarations.
13. Add the following code to the btnPrint_Click:

```
private async void btnPrint_Click(object sender, RoutedEventArgs
e)
{

    ReceiptPrintJob job = myClaimedPrinter.Receipt.CreateJob();
    job.PrintLine(txtInput.Text);
```

```
        job.PrintLine("");
        job.PrintLine("");
        job.PrintLine("");
        job.PrintLine("");
        job.PrintLine("");
        job.PrintLine("");
        job.CutPaper();
        if (!await job.ExecuteAsync())
        {
            txtStatus.Text = "Print failed";
        }
}
```

An instance of the ReceiptPrintJob class is used to print out the receipt. The job is to print the message from the textbox, print several blank lines to the printer so the message scrolls up the page, and then cut the paper. The job is then sent to the printer asynchronously.

14. Add the following code to the btnOpenCashDrawer_Click:

```
private    async    void    btnOpenCD_Click(object    sender,
RoutedEventArgs e)
{
    await myClaimedCD.OpenDrawerAsync();
}
```

The asynchronous method simply opens the cash drawer.

15. In the MainWindow() method, add the call to the POSSetup() method:

```
public MainWindow()
{
    InitializeComponent();
    POSSetup();
}
```

16. Remove any unused "using" declarations.
17. Save the whole project.

7.3.5   Test the Application
Now, we are ready to test the application.

1. In Visual Studio, build the application.
2. If there are any errors, correct them, and rebuild the application.

215

3. Make sure the printer is set up for the Ethernet interface, and the cash drawer is connected to the DK port.

4. Start the debugger. Once the application starts and a Pair Device box appears asking to link to the printer, click Allow.

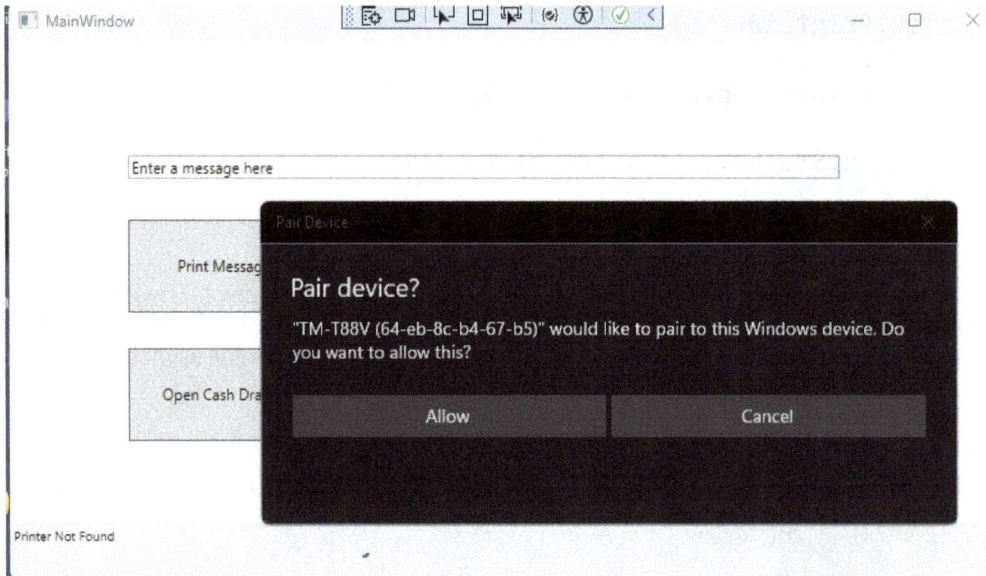

5. The claim for both devices will take place. The next box confirms the pairing. Click Close.

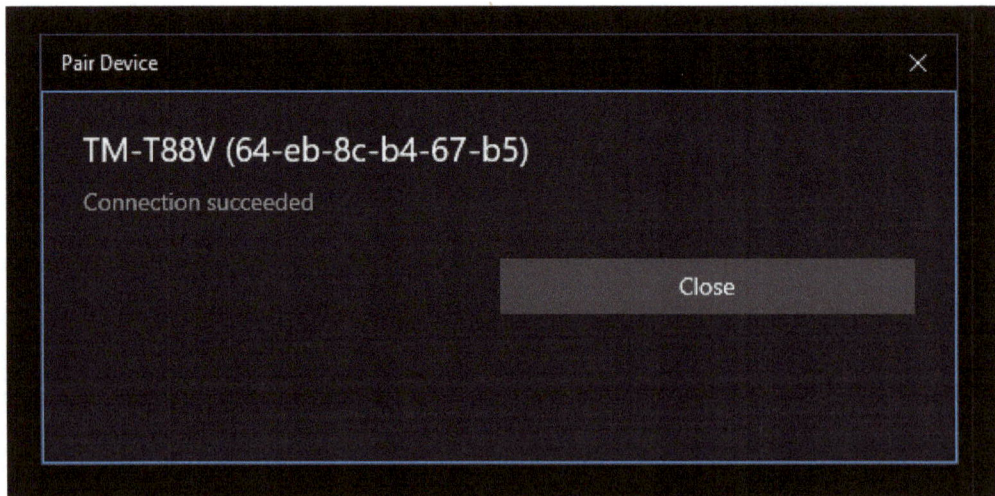

6. Enter some text in the text box and press the print button. The text should be printed out.
7. Press the open cash drawer button, and the cash drawer should open.
8. Open Device Manager, there is a new entry for the printer device:

> 🖥 Network adapters
∨ ⟨⟩ POS Remote Device
    ⟨⟩ TM-T88V (64-eb-8c-b4-67-b5)
> 🖨 Print queues

Check the properties, and you will see the RemotePosDrv.dll assembly is the driver.

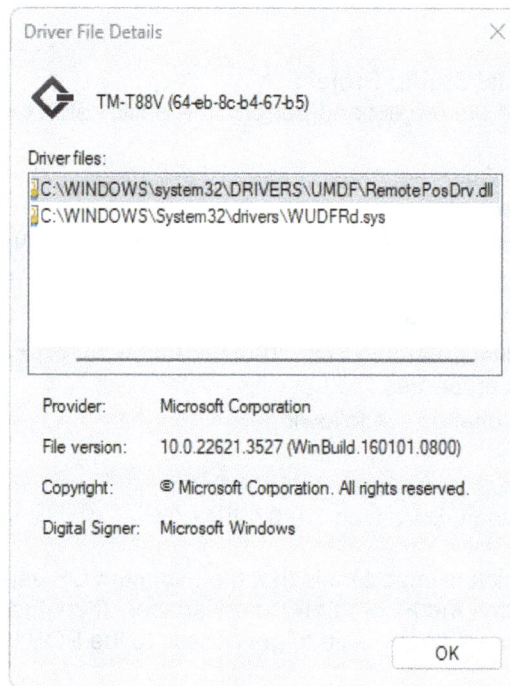

Driver File Details ✕

⟨⟩ TM-T88V (64-eb-8c-b4-67-b5)

Driver files:

C:\WINDOWS\system32\DRIVERS\UMDF\RemotePosDrv.dll
C:\WINDOWS\System32\drivers\WUDFRd.sys

Provider: Microsoft Corporation
File version: 10.0.22621.3527 (WinBuild.160101.0800)
Copyright: © Microsoft Corporation. All rights reserved.
Digital Signer: Microsoft Windows

OK

9. Turn off the printer, and then try to print something. The status box should say that the print failed.
10. Turn the printer back on, and then try to print something again. Turning on the printer will not recover the connection, because the remove event handler is never called for a network device. Either the application has to be restarted, or you will have to provide the user with the ability to perform a reconnection.
11. Stop debugging when finished.

7.4 Exercise 7.4: Webcam Barcode Scanner

A new feature added to POS WinRT is the ability to scan barcodes using a webcam. Webcams are a little more economical than the traditional barcode scanner, and they can easily be replaced. A production environment scanning items with a digital watermark moving on a conveyor belt is one example where a webcam scanner could operate. Another example is a small startup store, where a webcam is more budget-friendly. For a webcam to act as a barcode scanner, the webcam needs a quick autofocus feature to pick up the label. The webcam scanner supports the following barcode symbologies: PC/EAN, Code 39, Code 128, Interleaved 2 of 5, Databar Omnidirectional, Databar Stacked, QR Code, and GS1DWCode. In this exercise, we will create a webcam barcode scanner using the device watcher.

7.4.1 Create the Visual Studio Project
The first step is to create the project and perform the basic setup tasks.

1. Open Visual Studio.
2. Create a new project.
3. Search for and select "WPF Application for C#" and click Next.
4. Name the project EX74_Webcam and click Next.
5. Keep the .NET version at ".NET 8.0 (Long Term Support)" and click Create.
6. The next and most important step: from the menu, select Project-> EX74_Webcam properties.
7. Under General, change the following:

> Target OS Version: 10.0.26100.0
> Supported OS Version: 10.0.19041.0

The reason the OS version is important is that the minimum OS version of 10.0.19041.0 is the version that supports the POS WinRT namespaces. If you use any OS version below the minimum, you will not be able to get access to the POS WinRT namespaces.

8. Save the project when finished.
9. Close the EX74_Webcam properties page.

7.4.2 Set Up the XAML Controls
Now we will set up the UI.

1. In Solution Explorer, open the MainWindow.xaml.
2. Add a ListBox control, and resize to fit the top half of the page.
   a. Name: lstItems

        b.   Brush Foreground Color: Black
        c.   Brush Background Color: White
3.   Add the Status Strip
        a.   Name sBar
4.   Add a TextBlock control to the status strip.
        a.   Name: txtStatus

Here is the XAML code:

```xml
<Window x:Class="EX74_Webcam.MainWindow"

xmlns="http://schemas.microsoft.com/winfx/2006/xaml/presentatio
n"
        xmlns:x="http://schemas.microsoft.com/winfx/2006/xaml"

xmlns:d="http://schemas.microsoft.com/expression/blend/2008"
        xmlns:mc="http://schemas.openxmlformats.org/markup-
compatibility/2006"
        xmlns:local="clr-namespace:EX74_Webcam"
        mc:Ignorable="d"
        Title="MainWindow" Height="450" Width="800">
    <Grid>
        <ListBox x:Name="lstItems" d:ItemsSource="{d:SampleData
ItemCount=5}" Margin="0,10,0,159"/>
        <StatusBar x:Name="sBar" Margin="0,366,0,0">
            <TextBlock   x:Name="txtStatus"   TextWrapping="Wrap"
Text="Ready" Width="792" Height="54"/>
        </StatusBar>

    </Grid>
</Window>
```

The layout will look something like the following: (Since the border and capture elements are not visible, we left the Designer marks in the picture.)

```
MainWindow
Sample Item 1
Sample Item 2
Sample Item 3
Sample Item 4
Sample Item 5




Ready
```

5. Save the MainWindow.xaml file.

7.4.3 Write the code

The webcam scanner uses most of the same methods that we have been setting up for the barcode scanner. The difference will be a task for the capture element control. Let's create the code behind the interface.

1. Open the MainWindow.xaml.cs file.
2. Add the following using statements:

```
using Windows.Devices.PointOfService;
using Windows.Devices.Enumeration;
using Windows.System.Display;
using Windows.Storage.Streams;
```

The Windows.System.Display and Windows.Storage.Streams are for the webcam.

3. Above the MainWindow() method add the global instances for a BarcodeScanner, ClaimedBarcodeScanner, and Device Watcher.

```
public partial class MainWindow : Window
{

    BarcodeScanner scanner = null;
    ClaimedBarcodeScanner claimedScanner = null;
```

220

```
    DeviceWatcher                webCamDevWatcher              =
DeviceInformation.CreateWatcher(BarcodeScanner.GetDeviceSelecto
r(PosConnectionTypes.All));

    public MainWindow()
```

4. After the MainWindow() method, create the POSSetup() method:

```
private void POSSetup()
{
    webCamDevWatcher.Added += WebCamDevWatcher_Added;
    webCamDevWatcher.Removed += WebCamDevWatcher_Removed;
    webCamDevWatcher.Start();
}
```

5. Enter the code for the WebCamDevWatcher_Removed event handler:

```
private  void  WebCamDevWatcher_Removed(DeviceWatcher  sender,
DeviceInformationUpdate args)
{
    claimedScanner.Dispose();
    scanner.Dispose();
}
```

6. Enter the code for the WebCamDevWatcher_Added event handler:

```
private async void WebCamDevWatcher_Added(DeviceWatcher sender,
DeviceInformation args)
{
    scanner = await BarcodeScanner.FromIdAsync(args.Id);
    if (scanner != null)
    {

        claimedScanner = await scanner.ClaimScannerAsync();
        if (claimedScanner != null)
        {

            scanner.StatusUpdated += Scanner_StatusUpdated;
            claimedScanner.RetainDevice();
            claimedScanner.DataReceived                        +=
ClaimedScanner_DataReceived;
            claimedScanner.IsDecodeDataEnabled = true;
```

```
        await claimedScanner.EnableAsync();

        //bool            WebScannerSupportsPreview      =
!String.IsNullOrEmpty(scanner.VideoDeviceId);
        //if (WebScannerSupportsPreview)
        //{
        //    Code for starting the webcam preview
        //}

        //Start the capture
        await claimedScanner.StartSoftwareTriggerAsync();

        Application.Current.Dispatcher.Invoke((Action)(()
=> txtStatus.Text = "Barcode Scanner Claimed: " + args.Id));
        }
    else
    {
        Application.Current.Dispatcher.Invoke((Action)(()
=> txtStatus.Text = "Barcode Scanner Not Claimed"));
        }
    }
    else
    {
        Application.Current.Dispatcher.Invoke((Action)(()      =>
txtStatus.Text = "Barcode Scanner Not Found"));
    }
}
```

There are some differences from the other barcode programs. After the webcam has been claimed, rather than a separate event to retain the device, we simply make the call to RetainDevice(). All the other steps are the same. After the barcode has been enabled, the code checks to make sure we have a video device ID so we can show a web video preview. Since there is no CaptureElement control for WinUI or WPF like there is for UWP applications, a different solution is needed to show the video preview. For education purposes, the commented section has been left in the code for now so we can just show the solution works without the video preview. The final call is to the StartSoftwareTriggerAsync(); which starts the webcam to capture barcodes.

7. Next, fill in the code to the ClaimedScanner_DataReceived event handler:

```
private void ClaimedScanner_DataReceived(ClaimedBarcodeScanner
sender, BarcodeScannerDataReceivedEventArgs args)
{
```

222

```
    //Read the data from the buffer and convert to a string.
    var                      scanDataLabelReader                 =
DataReader.FromBuffer(args.Report.ScanDataLabel);
    string                     stringDataLabel                   =
scanDataLabelReader.ReadString(args.Report.ScanDataLabel.Length
);

    //Since this event is in a different thread, we need to sync
to the UI thread
    Dispatcher.BeginInvoke((Action)(()                          =>
lstItems.Items.Add(stringDataLabel)));
    Dispatcher.BeginInvoke((Action)(()  =>  txtStatus.Text  =
"Ready"));
}
```

8. Fill in the code for the Scanner_StatusUpdate event handler:

```
private  void  Scanner_StatusUpdated(BarcodeScanner  sender,
BarcodeScannerStatusUpdatedEventArgs args)
{
    uint bcEXStatus = args.ExtendedStatus;
    BarcodeScannerStatus bcStatus = args.Status;
    Dispatcher.Invoke((Action)(() => txtStatus.Text = "Barcode
status update: " + bcStatus.ToString()));
}
```

9. In the main method, add the call to the POSSetup() method:

```
public MainWindow()
{
    InitializeComponent();
    POSSetup();
}
```

10. Remove any unused "using" declarations.
11. Save the whole project.

7.4.4   Test the Application
Now, we are ready to test the application.

1. In Visual Studio, build the application.
2. If there are any errors, correct them, and rebuild the application.

3. If you are using a USB webcam, plug the webcam into your development machine.
4. Start the debugging of the application.
5. The webcam should be claimed.
6. Put barcodes in front of the webcam to scan. The webcam is not as fast as a regular barcode scanner, but it does pick up the barcode.

If you need some sample barcodes to test with the webcam, docs.microsoft.com has a page with example barcodes to test with a webcam scanner: https://docs.microsoft.com/en-us/windows/uwp/devices-sensors/pos-camerabarcode-symbologies. The picture below shows the output from some of the different symbologies from the webpage.

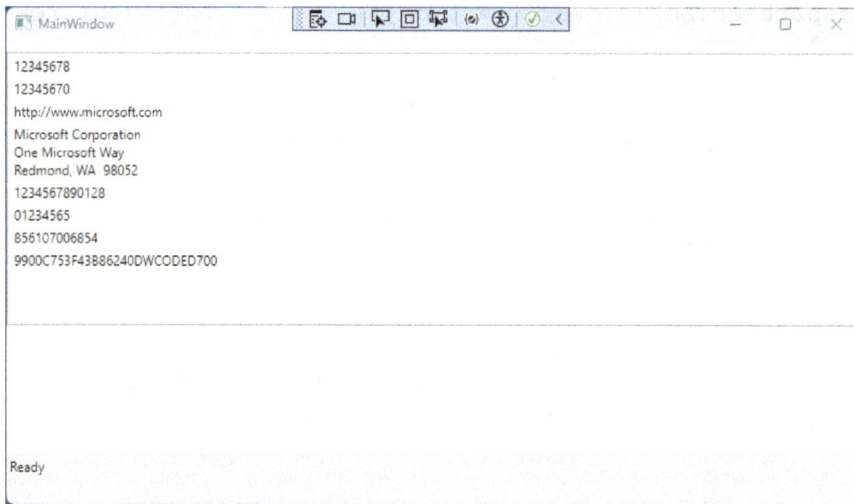

7. Stop the debugger when finished.

7.5 Exercise 7.5: Getting a List of All POS WinRT Devices using Snapshot

There are no POSDM.exe or SOManager utilities for POS WinRT devices, but you can get a list of devices that are connected to a Windows computer. We have seen the enumeration and connection to several different POS devices. Devices can exist locally (serial port or USB) or remotely (Ethernet or Bluetooth). Having users see a device name with a MAC address might not be user-friendly enough for most users. There might be multiple POS Printers on a network. Which one does the user select? For example, a restaurant might have a POS printer at each service station to generate customer receipts and a networked POS printer back in the kitchen to take orders. Knowing the name of each device can help to know what devices to connect to. In this exercise, we will use the Data Grid Control to store a list of POS devices attached to the Windows computer.

This exercise and the next are similar programs, but this one will use the snapshot solution. The snapshot solution is best used to create a custom picker list solution. The next project will use Device Watcher. The goal is to demonstrate the difference between the two enumeration types. These exercise examples serve as a base to create a more elaborate application.

7.5.1   Create the Visual Studio Project
The first step is to create the project and perform the basic setup tasks.

1.   Open Visual Studio.
2.   Create a new project.
3.   Search for and select "WPF Application for C#" and click Next.
4.   Name the project EX75_POSMGR_snapshot and click Next.
5.   Keep the .NET version at ".NET 8.0 (Long Term Support)" and click Create.
6.   The next and most important step: from the menu select Project-> EX75_POSMGR _snapshot properties.
7.   Under General, change the following:

> Target OS Version: 10.0.26100.0
> Supported OS Version: 10.0.19041.0

The reason the OS version is important is that the minimum OS version of 10.0.19041.0 is the version that supports the POS WinRT namespaces. If you use any OS version below the minimum, you will not be able to get access to the POS WinRT namespaces.

8.   Save the project when finished.
9.   Close the EX75_POSMGR _snapshot properties page.

7.5.2   Set Up the XAML Controls
Now we will set up the UI.

1.   In Solution Explorer, open the MainWindow.xaml.
2.   Add a Label control in the top left corner
     a.   VerticalAlignment: Top.
     b.   Name: lblTitleasd
     c.   Content: POS WinRT Devices
3.   Add a Datagrid below the label and leave room at the bottom for the status strip.
     a.   Name: posDeviceDG
4.   Add the Status Strip
     a.   Name sBar
     b.   VerticalAlignment: Bottom
5.   Add a TextBlock control to the status strip.

a. Name: txtStatus

Here is the XAML code:

```xml
<Window x:Class="EX75_ POSMGR_snapshot.MainWindow"

xmlns="http://schemas.microsoft.com/winfx/2006/xaml/presentatio
n"
        xmlns:x="http://schemas.microsoft.com/winfx/2006/xaml"

xmlns:d="http://schemas.microsoft.com/expression/blend/2008"
        xmlns:mc="http://schemas.openxmlformats.org/markup-
compatibility/2006"
        xmlns:local="clr-namespace:EX75_ POSMGR_snapshot"
        mc:Ignorable="d"
        Title="MainWindow" Height="450" Width="800">
    <Grid>
        <Label  x:Name="lblTitle"  Content="POS  WinRT  Devices"
Margin="28,18,609,0" FontSize="14" FontWeight="Bold" Height="26"
VerticalAlignment="Top"/>
        <DataGrid                              x:Name="posDeviceDG"
d:ItemsSource="{d:SampleData ItemCount=5}" Margin="0,49,0,49"/>
        <StatusBar Height="49" VerticalAlignment="Bottom">
            <TextBlock  x:Name="txtStatus"  TextWrapping="Wrap"
Text="Ready" Width="795" Height="36"/>
        </StatusBar>
    </Grid>
</Window>
```

The layout will look something like the following: (Since the border and capture elements are not visible, we left the Designer marks in the picture.)

6. Save the MainWindow.xaml file.

7.5.3   Write the code

Since we will only be gathering device information, the code doesn't have any device event handlers.

1. Open the MainWindow.xaml.cs file.
2. Add the following using statements:

```
using Windows.Devices.PointOfService;
using Windows.Devices.Enumeration;
using System.Data;
using System.Collections.ObjectModel;
```

3. Next, a DataTable will be used to help create a collection to be used as the DataGrid's source. Add a new DataTable before the MainWindows() method:

```
public partial class MainWindow : Window
{

    DataTable myDataTable = new DataTable();

    public MainWindow()
```

227

4. The next step is to create two methods that will set up the columns of the DataTable and another method to update the table with the devices found from each snapshot. Create a GridSetup() method after MainWindows() method:

```
public void GridSetup()
{
    myDataTable.Columns.Add("POS Type", typeof(string));
    myDataTable.Columns.Add("Device Name", typeof(string));
    myDataTable.Columns.Add("Device ID", typeof(string));
    myDataTable.Columns.Add("Kind", typeof(string));
    myDataTable.Columns.Add("IsDefault", typeof(bool));
    myDataTable.Columns.Add("IsEnabled", typeof(bool));

    //Add the columns
    for (int i = 0; i < myDataTable.Columns.Count; i++)
    {
        posDeviceDG.Columns.Add(new DataGridTextColumn()
        {
            Header = myDataTable.Columns[i].ColumnName,
            Binding = new Binding { Path = new PropertyPath("["
+ i.ToString() + "]") }
        });
    }
}
```

5. Create a GridUpdate() method after the GridSetup() method:

```
public void GridUpdate()
{
    //add the table data to the collection
    var collection = new ObservableCollection<object>();
    foreach (DataRow row in myDataTable.Rows)
    {
        collection.Add(row.ItemArray);
    }
    //add the collecton to the data grid
    posDeviceDG.ItemsSource = collection;
}
```

6. The last method to create is the POSSetup() method.

228

```
private async void POSSetup()
{

    txtStatus.Text = "Working. Please wait....";

    DeviceInformationCollection barcodeScanners = await
DeviceInformation.FindAllAsync(BarcodeScanner.GetDeviceSelector
(PosConnectionTypes.All));
    foreach (DeviceInformation devinfo in barcodeScanners)
    {
        myDataTable.Rows.Add("Barcode                      Scanner",
devinfo.Name.ToString(),                      devinfo.Id.ToString(),
devinfo.Kind.ToString(),          devinfo.IsDefault.ToString(),
devinfo.IsEnabled.ToString());
    }
    GridUpdate();


    txtStatus.Text = "Scanners Done!";

    DeviceInformationCollection magneticStripeReaders = await
DeviceInformation.FindAllAsync(MagneticStripeReader.GetDeviceSe
lector(PosConnectionTypes.All));
    foreach (DeviceInformation devinfo in magneticStripeReaders)
    {
        myDataTable.Rows.Add("MSR",        devinfo.Name.ToString(),
devinfo.Id.ToString(),                      devinfo.Kind.ToString(),
devinfo.IsDefault.ToString(), devinfo.IsEnabled.ToString());
    }

    GridUpdate();

    txtStatus.Text = "MSRs Done!";

    DeviceInformationCollection pollDisplays = await
DeviceInformation.FindAllAsync(LineDisplay.GetDeviceSelector(Po
sConnectionTypes.All));
    foreach (DeviceInformation devinfo in pollDisplays)
    {
        myDataTable.Rows.Add("Line                      Display",
devinfo.Name.ToString(),                      devinfo.Id.ToString(),
```

229

```
devinfo.Kind.ToString(),           devinfo.IsDefault.ToString(),
devinfo.IsEnabled.ToString());
    }

    GridUpdate();
    txtStatus.Text = "Line Displays Done!";

    DeviceInformationCollection      posPrinters      =      await
DeviceInformation.FindAllAsync(PosPrinter.GetDeviceSelector(Pos
ConnectionTypes.All));
    foreach (DeviceInformation devinfo in posPrinters)
    {
        myDataTable.Rows.Add("POS                          Printer",
devinfo.Name.ToString(),                   devinfo.Id.ToString(),
devinfo.Kind.ToString(),          devinfo.IsDefault.ToString(),
devinfo.IsEnabled.ToString());
    }
    GridUpdate();
    txtStatus.Text = "POS Printers Done!";

    DeviceInformationCollection      cashDrawers      =      await
DeviceInformation.FindAllAsync(CashDrawer.GetDeviceSelector(Pos
ConnectionTypes.All));
    foreach (DeviceInformation devinfo in cashDrawers)
    {
        myDataTable.Rows.Add("Cash                          Drawer",
devinfo.Name.ToString(),                   devinfo.Id.ToString(),
devinfo.Kind.ToString(),          devinfo.IsDefault.ToString(),
devinfo.IsEnabled.ToString());
    }
    GridUpdate();
    txtStatus.Text = "Cash Drawers Done!";

    txtStatus.Text = "";
}
```

The POSSetup() method makes a FindAllAsync() call for each POS WinRT supported device type. The code then calls GridUpdate to update the DataTable and the Datagrid with each device found.

7. The final step is to add the code to the MainWindows() method:

```
public MainWindow()
{
    InitializeComponent();
    GridSetup();
    POSSetup();
}
```

8. Save the project.

7.5.4   Test the Application
Now, we are ready to test the application.

1. In Visual Studio, build the application.
2. If there are any errors, correct them, and rebuild the application.
3. Make sure that you have all the devices connected and turned on so the application can find them.
4. Start the debugging of the application. The application will run and may take some time, but it will eventually list all the POS devices in the DataGrid.

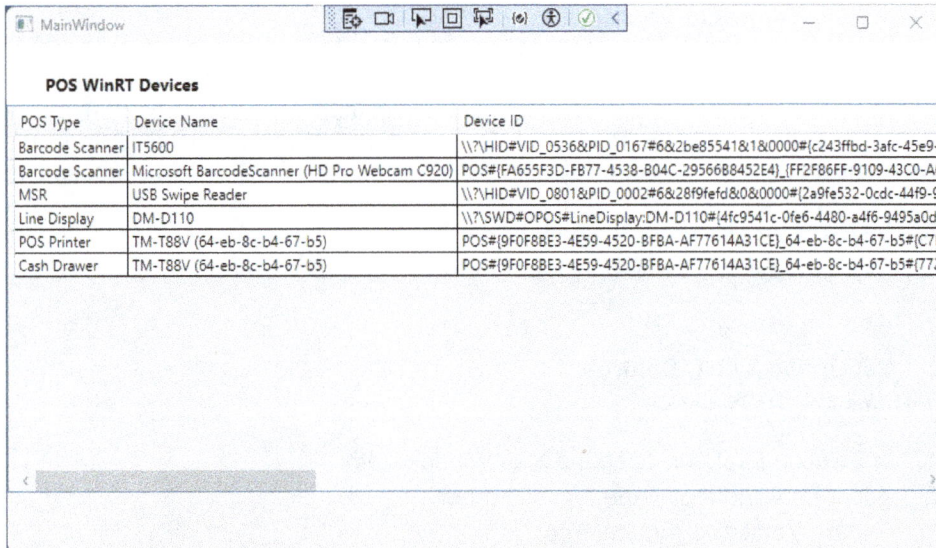

5. Stop debugging when finished.

7.6 Exercise 7.6: Getting a List of All POS WinRT Devices using Device Watcher

The snapshot enumeration method is fine for creating a list of devices to select from. If a USB device gets unplugged, the application doesn't dynamically update the DataGrid. For this exercise, we will create the same application, but use 5 device watchers to dynamically track devices with Added and Removed Events.

7.6.1 Create the Visual Studio Project

The first step is to create the project and perform the basic setup tasks.

1. Open Visual Studio.
2. Create a new project.
3. Search for and select "WPF Application for C#" and click Next.
4. Name the project EX76_POSMGR_devicewatcher and click Next.
5. Keep the .NET version at ".NET 8.0 (Long Term Support)" and click Create.
6. The next and most important step: from the menu select Project-> EX76_POSMGR_devicewatcher properties.
7. Under General, change the following:

> Target OS Version: 10.0.26100.0
> Supported OS Version: 10.0.19041.0

The reason the OS version is important is that the minimum OS version of 10.0.19041.0 is the version that supports the POS WinRT namespaces. If you use any OS version below the minimum, you will not be able to get access to the POS WinRT namespaces.

8. Save the project when finished.
9. Close the EX76_POSMGR_devicewatcher properties page.

7.6.2 Set Up the XAML Controls

Now, we will set up the UI.

1. In Solution Explorer, open the MainWindow.xaml.
2. Add a Label control in the top left corner
   a. VerticalAlignment: Top.
   b. Name: lblTitleasd
   c. Content: POS WinRT Devices
3. Add a Datagrid below the label and leave room at the bottom for the status strip
   a. Name: posDeviceDG
4. Add the Status Strip
   a. Name sBar

        b.   VerticalAlignment: Bottom
5.   Add a TextBlock control to the status strip.
        b.   Name: txtStatus

Here is the XAML code:

```xml
<Window x:Class="EX75_Webcam_snapshot.MainWindow"

xmlns="http://schemas.microsoft.com/winfx/2006/xaml/presentatio
n"
        xmlns:x="http://schemas.microsoft.com/winfx/2006/xaml"

xmlns:d="http://schemas.microsoft.com/expression/blend/2008"
        xmlns:mc="http://schemas.openxmlformats.org/markup-
compatibility/2006"
        xmlns:local="clr-namespace:EX75_Webcam_snapshot"
        mc:Ignorable="d"
        Title="MainWindow" Height="450" Width="800">
    <Grid>
        <Label   x:Name="lblTitle"   Content="POS   WinRT   Devices"
Margin="28,18,609,0" FontSize="14" FontWeight="Bold" Height="26"
VerticalAlignment="Top"/>
        <DataGrid                            x:Name="posDeviceDG"
d:ItemsSource="{d:SampleData ItemCount=5}" Margin="0,49,0,49"/>
        <StatusBar Height="49" VerticalAlignment="Bottom">
            <TextBlock   x:Name="txtStatus"   TextWrapping="Wrap"
Text="Ready" Width="795" Height="36"/>
        </StatusBar>
    </Grid>
</Window>
```

The layout will look something like the following: (Since the border and capture elements
are not visible, we left the Designer marks in the picture.)

MainWindow			
POS WinRT Devices			

SampleInt	SampleStringA	SampleStringB	SampleBool
1	Sample String A - 1	Sample String B - 1	☑
2	Sample String A - 2	Sample String B - 2	☑
3	Sample String A - 3	Sample String B - 3	☑
4	Sample String A - 4	Sample String B - 4	☑
5	Sample String A - 5	Sample String B - 5	☑

Ready

6. Save the MainWindow.xaml file.

7.6.3 Write the code
Since we will only be gathering device information, the code doesn't have any device event handlers.

1. Open the MainWindow.xaml.cs file.
2. Add the following using statements:

```
using Windows.Devices.PointOfService;
using Windows.Devices.Enumeration;
using System.Data;
using System.Collections.ObjectModel;
```

3. Next, a DataTable will be used to help create a collection to be used as the DataGrid's source. Add a new DataTable and all the device watchers before the MainWindows() method:

```
public partial class MainWindow : Window
{
    DeviceWatcher                BarCodeDevWatcher                =
DeviceInformation.CreateWatcher(BarcodeScanner.GetDeviceSelecto
r(PosConnectionTypes.All));
```

234

```
    DeviceWatcher              MSRDevWatcher              =
DeviceInformation.CreateWatcher(MagneticStripeReader.GetDeviceS
elector(PosConnectionTypes.All));
    DeviceWatcher            LineDisplayDevWatcher            =
DeviceInformation.CreateWatcher(LineDisplay.GetDeviceSelector(P
osConnectionTypes.All));
    DeviceWatcher            PosPrinterDevWatcher            =
DeviceInformation.CreateWatcher(PosPrinter.GetDeviceSelector(Po
sConnectionTypes.All));
    DeviceWatcher            CashdrawerDevWatcher            =
DeviceInformation.CreateWatcher(CashDrawer.GetDeviceSelector(Po
sConnectionTypes.All));
    DataTable myDataTable = new DataTable();

    public MainWindow()
```

4. The next step is to create two methods that will set up the columns of the
    DataTable and another method to update the table with the devices found
    from each snapshot. Create a GridSetup() method after MainWindows()
    method:

```
public void GridSetup()
{
    myDataTable.Columns.Add("POS Type", typeof(string));
    myDataTable.Columns.Add("Device Name", typeof(string));
    myDataTable.Columns.Add("Device ID", typeof(string));
    myDataTable.Columns.Add("Kind", typeof(string));
    myDataTable.Columns.Add("IsDefault", typeof(bool));
    myDataTable.Columns.Add("IsEnabled", typeof(bool));

    //Add the columns
    for (int i = 0; i < myDataTable.Columns.Count; i++)
    {
        posDeviceDG.Columns.Add(new DataGridTextColumn()
        {
            Header = myDataTable.Columns[i].ColumnName,
            Binding = new Binding { Path = new PropertyPath("["
+ i.ToString() + "]") }
        });
    }
}
```

5. Create a GridUpdate() method after the GridSetup() method:

```
public void GridUpdate()
{
    //add the table data to the collection
    var collection = new ObservableCollection<object>();
    foreach (DataRow row in myDataTable.Rows)
    {
        collection.Add(row.ItemArray);
    }
    //add the collecton to the data grid
    posDeviceDG.ItemsSource = collection;
}
```

6. The next method to create is the POSSetup() method, which will set up the device's watcher events and start each device watcher.

```
public void POSSetup()
{
    BarCodeDevWatcher.Added += BarCodeDevWatcher_Added;
    BarCodeDevWatcher.Removed += BarCodeDevWatcher_Removed;
    BarCodeDevWatcher.Start();

    MSRDevWatcher.Added += MSRDevWatcher_Added;
    MSRDevWatcher.Removed += MSRDevWatcher_Removed;
    MSRDevWatcher.Start();

    LineDisplayDevWatcher.Added += LineDisplayDevWatcher_Added;
    LineDisplayDevWatcher.Removed                            +=
LineDisplayDevWatcher_Removed;
    LineDisplayDevWatcher.Start();

    PosPrinterDevWatcher.Added += PosPrinterDevWatcher_Added;
    PosPrinterDevWatcher.Removed                            +=
PosPrinterDevWatcher_Removed;
    PosPrinterDevWatcher.Start();

    CashdrawerDevWatcher.Added += CashdrawerDevWatcher_Added;
    CashdrawerDevWatcher.Removed                            +=
CashdrawerDevWatcher_Removed;
    CashdrawerDevWatcher.Start();
```

```
}
```

7. Fill in the code for each of the Added and Removed event handlers:

```csharp
private void CashdrawerDevWatcher_Removed(DeviceWatcher sender,
DeviceInformationUpdate args)
{
    foreach (DataRow row in myDataTable.Rows)
    {
        if      (string.Equals(row["Device      ID"].ToString(),
args.Id.ToString(), StringComparison.OrdinalIgnoreCase))
        {
            myDataTable.Rows.Remove(row);
            break;
        }
    }
    Dispatcher.Invoke((Action)(() => GridUpdate()));
}

private void CashdrawerDevWatcher_Added(DeviceWatcher sender,
DeviceInformation args)
{
    myDataTable.Rows.Add("Cash   Drawer",   args.Name.ToString(),
args.Id.ToString(),                    args.Kind.ToString(),
args.IsDefault.ToString(), args.IsEnabled.ToString());
    Dispatcher.Invoke((Action)(() => GridUpdate()));
}
private void PosPrinterDevWatcher_Removed(DeviceWatcher sender,
DeviceInformationUpdate args)
{
    foreach (DataRow row in myDataTable.Rows)
    {

        if      (string.Equals(row["Device      ID"].ToString(),
args.Id.ToString(), StringComparison.OrdinalIgnoreCase))
        {
            myDataTable.Rows.Remove(row);
            break;
        }
    }
    Dispatcher.Invoke((Action)(() => GridUpdate()));
}
```

```
private void PosPrinterDevWatcher_Added(DeviceWatcher sender,
DeviceInformation args)
{
    myDataTable.Rows.Add("POS    Printer",    args.Name.ToString(),
args.Id.ToString(),                      args.Kind.ToString(),
args.IsDefault.ToString(), args.IsEnabled.ToString());
    Dispatcher.Invoke((Action)(() => GridUpdate()));
}
private void LineDisplayDevWatcher_Removed(DeviceWatcher sender,
DeviceInformationUpdate args)
{
    foreach (DataRow row in myDataTable.Rows)
    {

        if      (string.Equals(row["Device      ID"].ToString(),
args.Id.ToString(), StringComparison.OrdinalIgnoreCase))
        {
            myDataTable.Rows.Remove(row);
            break;
        }
    }
    Dispatcher.Invoke((Action)(() => GridUpdate()));
}

private void LineDisplayDevWatcher_Added(DeviceWatcher sender,
DeviceInformation args)
{
    myDataTable.Rows.Add("Line    Display",   args.Name.ToString(),
args.Id.ToString(),                      args.Kind.ToString(),
args.IsDefault.ToString(), args.IsEnabled.ToString());
    Dispatcher.Invoke((Action)(() => GridUpdate()));
}

private    void    MSRDevWatcher_Removed(DeviceWatcher    sender,
DeviceInformationUpdate args)
{
    foreach (DataRow row in myDataTable.Rows)
    {

        if      (string.Equals(row["Device      ID"].ToString(),
args.Id.ToString(), StringComparison.OrdinalIgnoreCase))
        {
```

```csharp
            myDataTable.Rows.Remove(row);
            break;
        }
    }
    Dispatcher.Invoke((Action)(() => GridUpdate()));
}


private void MSRDevWatcher_Added(DeviceWatcher sender,
DeviceInformation args)
{
    myDataTable.Rows.Add("MSR", args.Name.ToString(),
args.Id.ToString(), args.Kind.ToString(),
args.IsDefault.ToString(), args.IsEnabled.ToString());
    Dispatcher.Invoke((Action)(() => GridUpdate()));
}


private void BarCodeDevWatcher_Removed(DeviceWatcher sender,
DeviceInformationUpdate args)
{
    foreach (DataRow row in myDataTable.Rows)
    {

        if (string.Equals(row["Device ID"].ToString(),
args.Id.ToString(), StringComparison.OrdinalIgnoreCase))
        {
            myDataTable.Rows.Remove(row);
            break;
        }
    }
    Dispatcher.Invoke((Action)(() => GridUpdate()));
}


private void BarCodeDevWatcher_Added(DeviceWatcher sender,
DeviceInformation args)
{
    myDataTable.Rows.Add("Barcode Scanner",
args.Name.ToString(), args.Id.ToString(), args.Kind.ToString(),
args.IsDefault.ToString(), args.IsEnabled.ToString());
    Dispatcher.Invoke((Action)(() => GridUpdate()));
}
```

There is more code for this version of the application. The application will update the DataGrid when there is a device change. Each Added and Removed event pair is similar. If a device is found or added, it is added to the DataTable and DataGrid. If a device is removed, a search is performed in the DataTable based on the deviceID, the row is removed, and the DataGrid is then updated. There are probably more efficient data table queries that could be performed for larger data tables, but the application demonstrates the concept.

8. Finally, add the code in the MainWindow() method:

```
public MainWindow()
{
    InitializeComponent();
    GridSetup();
    POSSetup();
}
```

9. Save the project.

7.6.4   Test the Application

Now, we are ready to test the application.

1. In Visual Studio, build the application.
2. If there are any errors, correct them, and rebuild the application.
3. Make sure that you have all the devices connected and turned on so the application can find them.
4. Start the debugging of the application. This version of the application lists all the devices much faster than the snapshot method.
5. Unplug any connected USB webcam, USB barcode scanner, or USB MSR. The DataGrid gets updated immediately.
6. Stop debugging when finished.

Note: There is no code for the status strip, but the application could be expanded to provide user feedback.

7.7 *Summary*

The chapter covered the remaining four POS WinRT devices. Each device has its own unique setup process. The Line Display required the 3rd party OPOS driver to be set up before the application could connect to the device. Remote devices connected via Ethernet or Bluetooth require pairing between the device and Windows computer to complete the connection. There are 31 more devices in the Unified Specification. The next chapter will demonstrate how to access the remaining 31 devices using a class library.

8 POS WinRT and POS for .NET Working Together

The last chapter covered five supported POS WinRT devices. The big question is: what about the other 31 POS devices called out in the UnifiedPOS specification? There was a reason why POS for .NET was covered first because POS for .NET is the answer to that question. This chapter will cover a solution to access POS scale devices from a WPF .NET application. The POS scale Service Object created in Chapter 5 will be used in this example. The chapter will cover the following:

- Backstory of UWP and bridge technologies
- Using named pipes
- Demonstration of a solution for the POS Scale

Note: If you don't have the scale, you can subtitle a different POS device as the device to be accessed using POS for .NET.

8.1 A Little Backstory

Chapter 1 covered the backstory of UWP, but there is a little more to the story. The good folks at Microsoft recognized that UWP was not getting as much traction as envisioned. UWP applications are very siloed and limit access compared to desktop applications. The restrictions were so severe that developers were not buying into the concept. Once UWP matured to a certain point, they went back to see what could be done to help Win32 developers migrate. The result was the Desktop Bridge program. Several examples were created to help developers migrate their current applications to UWP or provide a mechanism to allow desktop applications to be available on the Microsoft Store. One of the solutions is the ability to call Windows runtime (RT) APIs from a traditional desktop application, and within that solution is the ability to create a console application service that can communicate with a UWP application. The book *Professional's Guide to POS for Windows Runtime* demonstrated this solution, which turned out to be quite complicated.

Microsoft realized UWP was dead, and they slowly moved support to WinUI, MAUI, and .NET. The bridging technologies are no longer needed, but the concept of a .NET UI

243

application communicating with a .NET Framework console application is still a solution for accessing the remaining 31 support POS devices.

8.2   What Cannot be Done

A WPF .NET Framework application using POS for .NET was created in Chapter 3. If POS for .NET is added as a reference to a .NET application, a run-time exception error will be triggered when the POS Explorer instance is created:

```
System.TypeInitializationException: 'The type initializer for
'Microsoft.PointOfService.Management.Explorer' threw an exception.'

MissingMethodException: Method not found: 'Void
System.AppDomainSetup.set_ApplicationBase(System.String)'.
```

.NET applications can access .NET Framework class libraries, but when it comes to POS for .NET, the same error occurs. Here is an example of a class library for the POS scale that was intended for a .NET application to access:

```csharp
using System;
using System.Collections.Generic;
using System.Linq;
using System.Net.NetworkInformation;
using System.Text;
using System.Threading.Tasks;
using Microsoft.PointOfService;

namespace ABscale
{
    public class ABscale6710
    {

        static private PosExplorer localPosExplorer;
        static private Scale localScale;

        public bool ABscaleSetup()
        {

            localPosExplorer = new PosExplorer();
            DeviceInfo                    device                    =
localPosExplorer.GetDevice("Scale", "ABScale");

            if (device == null)
```

244

```
            {
                return false;
            }
            else
            {
                localScale                                    =
(Scale)localPosExplorer.CreateInstance(device);
                localScale.Open();
                localScale.Claim(1000);
                localScale.DeviceEnabled = true;
                return true;
            }
        }

    public string GetWeight()
    {
        Decimal theWeight = 0;
        String ScaleStatus;

        try
        {
            theWeight = localScale.ReadWeight(1000);
        }
        catch
        {
            theWeight = 0;
            return "Error";
        }

        if (theWeight.ToString() == "-1")
        {

            ScaleStatus                                    =
localScale.CheckHealth(HealthCheckLevel.Internal);
            switch (ScaleStatus)
            {
                case "11":
                    return "Error 11";

                case "12":
                    return "Error 12";
```

```
                case "13":
                    return "Error 13";

                default:
                    return "Unkown Error";
            }
        }
        else
        {
            return theWeight.ToString() + " lbs";
        }
    }

    public bool ZeroScale()
    {
        localScale.ZeroScale();
        return true;
    }

    public string GetStatus()
    {
        String              ScaleStatus              =
localScale.CheckHealth(HealthCheckLevel.Internal);
        switch (ScaleStatus)
        {
            case "11":
                return "Scale Stable and not at Zero";

            case "12":
                return "Scale is in motion and not ready";

            case "13":
                return "Scale is at Zero and Stable";

            default:
                return "Unkown Error";
        }
    }

}
}
```

The class library compiles; but when a .NET application attempts to use the library, the same runtime error will be triggered upon the creation of the POS Explorer instance. POS for .NET expects to be running in its own application process space. POS for .NET is older than .NET Core, and it was not designed for the invocation calls from a different runtime. Per UWP bridge solution, the concept of using a separate .NET Framework application to run POS for .NET is the best solution. The next step is how to communicate between the .NET and .NET Framework applications.

8.3   Inter-Process Communication: Pipes

Inter-process communication is needed for the two applications to talk to each other. C# offers different inter-process communication solutions: shared memory buffer, memory mapped file, pipes, sockets, etc. The solution used to access the POS scale device will be through pipes. There are two different pipe types. The first is anonymous pipes, which are intended for a local computer and provide one-way communication. The second is named pipes, which can be networked and allow for two-way communication. Since the two applications will send data back and forth, named pipes will be used for inter-process communication.

8.4   Exercise 8.1: POS Scale Server in POS for .NET

The full solution and test will be broken into three separate exercises. In this exercise, we will create the .NET Framework console application to handle the POS for .NET and POS Scale calls for getting the weight, zeroing the scale, and checking the scale health.

8.4.1   Part 1: Create the Application
The .NET Framework console application will act as a server for the WPF .NET to access.

1. Open Visual Studio.
2. Click on "Create a new Project".
3. Set the language to C# and search for "Class Library (.NET Framework) template.

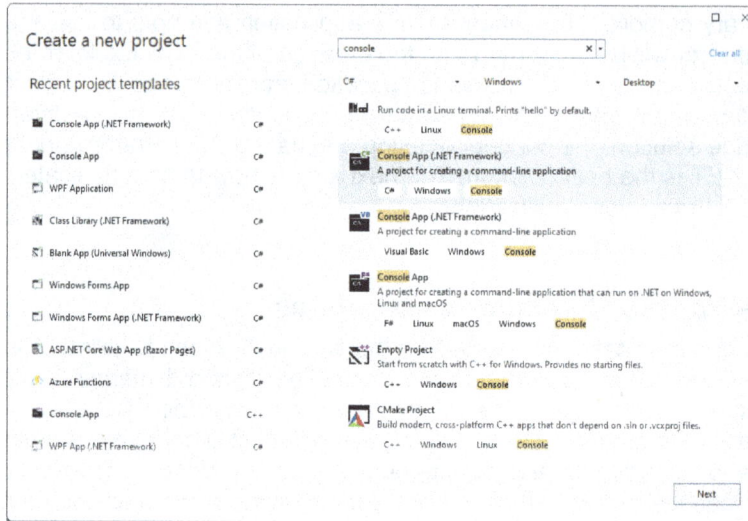

4. Click Next.
5. Name the project ABScaleServer.
6. Set the location for the project.
7. Keep the Framework set to .NET Framework 4.8.
8. Click Create.

8.4.2 Part 2: Adding the POS for .NET Libraries and Code
The next step is to add the POS for .NET Library and fill in the code.

1. From the menu, select Project->Add Reference. This will open the Add Reference dialog.
2. Click on the Browse tab, and locate the Microsoft.PointOfService.dll found under "C:\Program Files(x86)\Microsoft point of Service\SDK".
3. Click on the Add button.

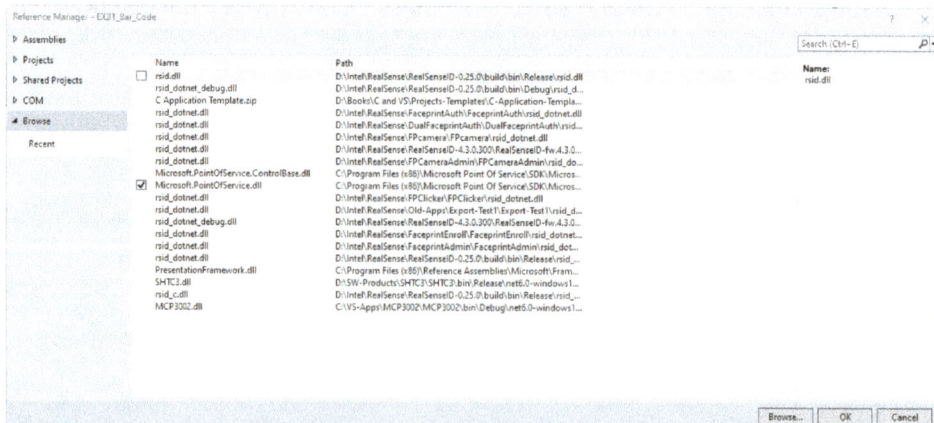

4. With the Microsoft.PointOfService.dll selected, click OK. The Microsoft.PointOfService reference is added to the project.
5. Add the following using statements to add support for POS for .NET and Pipes:

```
using System.Threading.Tasks;
using System.IO;
using System.IO.Pipes;
using Microsoft.PointOfService;
```

6. Before the Main() method, add the globals for PosExplorer and the Scale:

```
namespace ABScaleServer
{
    internal class Program
    {

        static private PosExplorer localPosExplorer;
        static private Scale localScale;

        static async Task Main(string[] args)
```

7. All the code is going to go into the Main() method. Since there are going to be asynchronous calls within the method and the WPF application is going to be making asynchronous calls, the Main() method will be defined with async and Task. Enter the following code, and the explanation will come after:

```
static async Task Main(string[] args)
{

    localPosExplorer = new PosExplorer();
    DeviceInfo device = localPosExplorer.GetDevice("Scale",
"ABScale");

    try
    {
        localScale                                           =
(Scale)localPosExplorer.CreateInstance(device);
        localScale.Open();
        localScale.Claim(1000);
        localScale.DeviceEnabled = true;
    }
    catch (Exception ex)
```

249

```csharp
    {

        throw ex;
    }

    if (localScale.DeviceEnabled)
    {

        var          pipeScaleServer              =          new
NamedPipeServerStream("scalepipe", PipeDirection.InOut);

        StreamReader reader = new StreamReader(pipeScaleServer);
        StreamWriter writer = new StreamWriter(pipeScaleServer);
        Console.WriteLine("Waiting for connection");

        do
        {
            try
            {
                pipeScaleServer.WaitForConnection();
                Console.WriteLine("Connection");
                string          posScalecommand          =          await
reader.ReadLineAsync();

                if (posScalecommand.Equals("getweight"))
                {
                    string theweight = "0.00";
                    try
                    {
                        theweight                                    =
localScale.ReadWeight(1000).ToString();
                    }
                    catch (Exception ex) {
                        theweight = "Error";
                    }
                    Console.WriteLine("Reading   weight:   "   +
theweight);
                    await writer.WriteLineAsync(theweight);
                    await writer.FlushAsync();
                    pipeScaleServer.WaitForPipeDrain();
                }
                if (posScalecommand.Equals("zero"))
```

```
                {
                    Console.WriteLine("Zeroing scale");
                    localScale.ZeroScale();
                }
                if (posScalecommand.Equals("health"))
                {
                    string              thehealth               =
localScale.CheckHealth(HealthCheckLevel.Internal).ToString();

                    switch (thehealth)
                    {
                        case "11":
                            await   writer.WriteLineAsync("Scale
Stable and not at Zero");
                            await writer.FlushAsync();
                            break;

                        case "12":
                            await   writer.WriteLineAsync("Scale
is in motion and not ready");
                            await writer.FlushAsync();
                            break;

                        case "13":
                            await   writer.WriteLineAsync("Scale
is at Zero and Stable");
                            await writer.FlushAsync();
                            break;

                        default:
                            await   writer.WriteLineAsync("Unkown
Error");
                            await writer.FlushAsync();
                            break;
                    }
                }

                if (posScalecommand.Equals("close"))
                {
                    Console.WriteLine("Exiting");
                    break;
                }
```

```
        }
        catch (Exception ex)
        {
            throw ex;
        }
        finally
        {
            pipeScaleServer.WaitForPipeDrain();
            if (pipeScaleServer.IsConnected)
            {
                pipeScaleServer.Disconnect();
            }
        }
    } while (true);
}
}
```

The first two calls in Main() are to instantiate PosExplorer and get the device information for the POS scale devices. The logical name is what we gave the POS Scale in Chapter 5.

The first try-catch attempts to open and claim the POS Scale. If successful, DeviceEnabled is set to True. If there was a failure to open and claim the scale, the code would throw an exception and close.

If DeviceEnabled is true, the big if statement contains the main body of the code. The NamePipeServerStream is set up with a name, "scalepipe", and two-way communication. StreamReader and StreamWriter instances are created. Being a console application, a message is sent out to indicate everything is ready for communication.

Next comes the Do-While loop with a try-catch-finally inside. The console sits and waits for a connection. Once a connection is made, a string is read into the posScalecommand. Based on the string sent, several if statements will process the command:

- "getweight" - the POS for .NET ReadWeight call is made and the value is written to the pipe.
- "zero" - the POS for .NET call to zero the scale is made and nothing is sent back through the pipe.
- "health" – a string that aligns with the scale return code is sent back through the pipe.
- "close" – a signal from the WPF application that the application is closed and it should also close.
- The break will jump out of the do-while loop and end the console application.

An exception is thrown if there are any errors. The Finally statement does the cleanup of the named pipe by draining and closing the pipe so the do-while loop can run again and wait for a connection to the pipe.

8. Build the application
9. Correct any errors and re-build.

8.5   Exercise 8.2: Barcode and Weight Scale in POS WinRT

The next step is to create the GUI application that will talk to the POS Scale console application. The application will support a barcode scanner and a weight scale. The barcode scanner will use the WinRT calls and the weight scale will make calls to the POS Scale console application. The concept for the project is to create an application that reads a product's barcode and records the shipping weight in a database. For this example, we will focus only on the barcode scanner and scale functionality.

8.5.1   Create the Visual Studio Project
The first step is to create the project and perform the basic setup tasks.

1. Open Visual Studio.
2. Create a new project.
3. Search for and select "WPF Application for C#" and click Next.
4. Name the project EX82_Barcode_Scale and click Next.
5. Keep the .NET version at ".NET 8.0 (Long Term Support)" and click Create.
6. The next and most important step: from the menu select Project-> EX82_Barcode_Scale properties.
7. Under General, change the following:

> Target OS Version: 10.0.26100.0
> Supported OS Version: 10.0.19041.0

The reason the OS version is important is that the minimum OS version of 10.0.19041.0 is the version that supports the POS WinRT namespaces. If you use any OS version below the minimum, you will not be able to get access to the POS WinRT namespaces.

8. Save the project when finished.
9. Close the EX82_Barcode_Scale properties page.

8.5.2   Set Up the XAML Controls
Now we will set up the UI.

1. In Solution Explorer, open the MainPage.xaml.
2. Add a ListBox under the label and resize the LisBox to fill 1/2 of the page.

253

a. Name: lstItems
3. Add a Button below the listbox.
    a. Name: btnWeight
    b. Content: Get Weight
    c. Font Size: 20
4. Add a Label next to the button.
    a. Name: lblWeight
    b. Content: Weight:
    c. Font Size: 20
5. Add a TextBlock next to the label.
    a. Name: theWeight
    b. Text: 0.00
    c. Font Size: 20
6. Add a Button to the right of the TextBlock.
    a. Name: btnZero
    b. Content: Zero
    c. Font Size: 20
7. Add a Button below the btnZero button.
    a. Name: btnScaleChekc
    b. Content: Scale Status
    c. Font Size: 20
8. Add a StatusBar at the bottom of the page.
    a. Name: sBAR
9. Add a TextBlock control to the status strip.
    a. Name: txtStatus
    b. Text: Ready

Here is the XAML code:

```
<Window x:Class="EX77_barcode_scale.MainWindow"

xmlns="http://schemas.microsoft.com/winfx/2006/xaml/presentatio
n"
        xmlns:x="http://schemas.microsoft.com/winfx/2006/xaml"

xmlns:d="http://schemas.microsoft.com/expression/blend/2008"
        xmlns:mc="http://schemas.openxmlformats.org/markup-
compatibility/2006"
        xmlns:local="clr-namespace:EX77_barcode_scale"
        mc:Ignorable="d"
        Title="MainWindow" Height="450" Width="800">
    <Grid>
```

```xml
        <ListBox x:Name="lstItems" d:ItemsSource="{d:SampleData
ItemCount=5}" Margin="0,0,0,178"/>
        <StatusBar x:Name="sBar" Margin="0,379,0,0">
            <TextBlock x:Name="txtStatus" TextWrapping="Wrap"
Text="Ready" Width="795" Height="41"/>
        </StatusBar>
        <TextBlock x:Name="txtWeight" HorizontalAlignment="Left"
Margin="298,285,0,0"         TextWrapping="Wrap"         Text="0.00"
VerticalAlignment="Top" Width="233" Height="43" FontSize="20"/>
        <Label         x:Name="lblWeight"         Content="Weight:"
HorizontalAlignment="Left"              Margin="199,280,0,0"
VerticalAlignment="Top" FontSize="20"/>
        <Button     x:Name="btnWeight"     Content="Get     Weight"
HorizontalAlignment="Left"              Margin="21,280,0,0"
VerticalAlignment="Top"  Width="154"  Height="32"  FontSize="20"
Click="btnWeight_Click"/>
        <Button      x:Name="btnZero"      Content="Zero      Scale"
HorizontalAlignment="Left"              Margin="613,274,0,0"
VerticalAlignment="Top"  Height="32"  Width="154"  FontSize="20"
Click="btnZero_Click"/>
        <Button  x:Name="btnScaleCheck"  Content="Scale  Status"
HorizontalAlignment="Left"              Margin="613,326,0,0"
VerticalAlignment="Top"  Height="30"  Width="154"  FontSize="20"
Click="btnScaleCheck_Click"/>

    </Grid>
</Window>
```

The Designer should show the following:

255

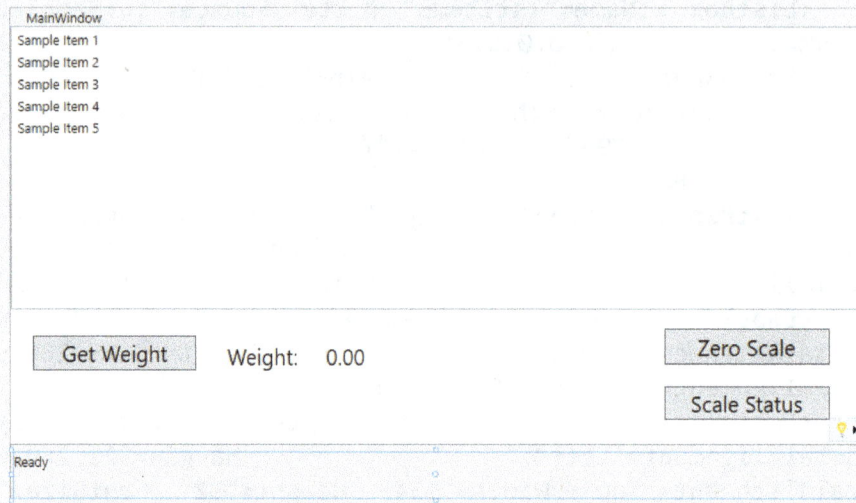

10. Save the MainPage.xaml file.

8.5.3   Write the code

Now, we will create the code behind the interface.

1. Open the MainPage.xaml.cs file.
2. Add the following using statements so we can access the PointOfService namespace enumeration namespace, and Pipes:

```
using Windows.Devices.PointOfService;
using Windows.Devices.Enumeration;
using Windows.Storage.Streams;
using System.IO;
using System.IO.Pipes;
using System.Diagnostics;
```

3. Inside the Class and above the MainWindow() method, add the following globals:

```
BarcodeScanner scanner = null;
ClaimedBarcodeScanner claimedScanner = null;
DeviceWatcher                posDevWatcher                =
DeviceInformation.CreateWatcher(BarcodeScanner.GetDeviceSelecto
r(PosConnectionTypes.All));
```

4. Below the MainWindow() method, add the POSSetup() method:

256

```
private void POSSetup()
{
    posDevWatcher.Added += PosDevWatcher_Added;
    posDevWatcher.Removed += PosDevWatcher_Removed;
    posDevWatcher.Start();
}
```

5. Fill in the code for the PosDevWatcher_Removed() event handler:

```
private   void   PosDevWatcher_Removed(DeviceWatcher   sender,
DeviceInformationUpdate args)
{
    claimedScanner.DataReceived -= ClaimedScanner_DataReceived;
    claimedScanner.ReleaseDeviceRequested                        -=
ClaimedScanner_ReleaseDeviceRequested;
    scanner.StatusUpdated -= Scanner_StatusUpdated;
    claimedScanner.Dispose();
    scanner.Dispose();
    Application.Current.Dispatcher.Invoke((Action)(()            =>
txtStatus.Text = "Barcode Scanner Removed"));
}
```

6. Fill in the code for the PosDevWatcher_Added event handler:

```
private  async  void  PosDevWatcher_Added(DeviceWatcher  sender,
DeviceInformation args)
{
    scanner = await BarcodeScanner.FromIdAsync(args.Id);
    if (scanner != null)
    {

        claimedScanner = await scanner.ClaimScannerAsync();
        if (claimedScanner != null)
        {
            scanner.StatusUpdated += Scanner_StatusUpdated;
            claimedScanner.ReleaseDeviceRequested               +=
ClaimedScanner_ReleaseDeviceRequested;
            claimedScanner.DataReceived                         +=
ClaimedScanner_DataReceived;
            claimedScanner.IsDecodeDataEnabled = true;
            await claimedScanner.EnableAsync();
```

```
            Application.Current.Dispatcher.Invoke((Action)(()
=> txtStatus.Text = "Barcode Scanner Claimed: " + args.Id));

        }
        else
        {
            Application.Current.Dispatcher.Invoke((Action)(()
=> txtStatus.Text = "Barcode Scanner Not Claimed"));
        }

    }
    else
    {
        Application.Current.Dispatcher.Invoke((Action)(()      =>
txtStatus.Text = "Barcode Scanner Not Found"));
    }
}
```

7. Fill in the code for the ClaimedScanner_DataReceived event handler:

```
private void ClaimedScanner_DataReceived(ClaimedBarcodeScanner
sender, BarcodeScannerDataReceivedEventArgs args)
{
    //Read the data from the buffer and convert to a string.
    var                    scanDataLabelReader               =
DataReader.FromBuffer(args.Report.ScanDataLabel);
    string                 stringDataLabel                   =
scanDataLabelReader.ReadString(args.Report.ScanDataLabel.Length
);

    //Since this event is in a different thread, we need to sync
to the UI thread
    Dispatcher.BeginInvoke((Action)(()                         =>
lstItems.Items.Add(stringDataLabel)));
    Dispatcher.BeginInvoke((Action)(()    =>   txtStatus.Text  =
"Ready"));
}
```

8. Fill in the code for ClaimedScanner_ReleaseDeviceRequested event handler:

```
private     void     ClaimedScanner_ReleaseDeviceRequested(object?
sender, ClaimedBarcodeScanner e)
{
    Dispatcher.BeginInvoke((Action)(()   =>   txtStatus.Text   =
"Barcode being released"));
}
```

9. Fill in the code for Scanner_StatusUpdated event handler:

```
private    void    Scanner_StatusUpdated(BarcodeScanner    sender,
BarcodeScannerStatusUpdatedEventArgs args)
{
    uint bcEXStatus = args.ExtendedStatus;
    BarcodeScannerStatus bcStatus = args.Status;
    Dispatcher.Invoke((Action)(() => txtStatus.Text = "Barcode
status update: " + bcStatus.ToString()));
}
```

10. All of the code for the barcode scanner is the same as the Barcode Device Watcher application from Chapter 6. The next step is to add the code for the buttons. Open MainWindow.xaml.
11. Double-click on all three buttons to generate the event handlers in the code.
12. In MainWindows.xaml.cs, add the following code for the btnWeight_Click event handler:

```
private async void btnWeight_Click(object sender, RoutedEventArgs e)
{
  try
  {
    var pipeClient = new NamedPipeClientStream(".", "scalepipe",
PipeDirection.InOut);
    pipeClient.Connect();

    var reader = new StreamReader(pipeClient);
    var writer = new StreamWriter(pipeClient);

    await writer.WriteLineAsync("getweight");
    await writer.FlushAsync();

    txtWeight.Text = await reader.ReadToEndAsync();
    pipeClient.Close();
  }
```

```
catch(Exception ex) {

    txtStatus.Text = ex.Message;
  }
}
```

The code creates an instance of the pipe client for a local computer to access the pipe named "scalepipe". Like the server applications, the communication is two-way. A connection is made. A StreamReader and StreamWriter instances are created. The command to "getweight" is sent to the server. The code waits for a response from the server before closing the pipe.

13. Fill in the code for the btnZero_Click event handler:

```
private async void btnZero_Click(object sender, RoutedEventArgs
e)
{
    try
    {
        var   pipeClient   =   new   NamedPipeClientStream(".",
"scalepipe", PipeDirection.InOut);
        pipeClient.Connect();

        var reader = new StreamReader(pipeClient);
        var writer = new StreamWriter(pipeClient);

        await writer.WriteLineAsync("zero");
        await writer.FlushAsync();
        pipeClient.Close();
    }
    catch (Exception ex) {

        txtStatus.Text = ex.Message;
    }
}
```

Like the call to get the weight, the connection to the local pipe is made, and the command to zero the scale is sent. There is nothing to read back, so the pipe is closed. Of course, a return message is just good practice for completeness.

14. Fill in the code for the btnScaleCheck_Click event handler:

```
private    async    void    btnScaleCheck_Click(object    sender,
RoutedEventArgs e)
{

    try
    {
        var    pipeClient    =    new    NamedPipeClientStream(".",
"scalepipe", PipeDirection.InOut);
        pipeClient.Connect();

        var reader = new StreamReader(pipeClient);
        var writer = new StreamWriter(pipeClient);

        await writer.WriteLineAsync("health");
        await writer.FlushAsync();

        txtStatus.Text = await reader.ReadToEndAsync();

        pipeClient.Close();

    }
    catch (Exception ex) {
        txtStatus.Text = ex.Message;
    }
}
```

Again, the same steps to connect to the local pipe, send the command, and wait for a response are made.

15. Finally, add the POSSetup() call to MainWindow():

```
public MainWindow()
{
    InitializeComponent();
    POSSetup();
}
```

16. Save the project.

8.6 Exercise 8.3: Testing Both Applications

Two instances of Visual Studio will be running for this test.

1. Plug the barcode scanner into the computer.
2. Connect the Avery Berkel 6710 scale to the computer via a serial or USB-to-serial connection.
3. Make sure the ABScale Service Object is set up per Section 5.4.3. Double-check the Service Object path and COM port setting.
4. Open Visual Studio and open the ABScaleServer project.
5. Start debugging the application. The application starts and waits for a connection.

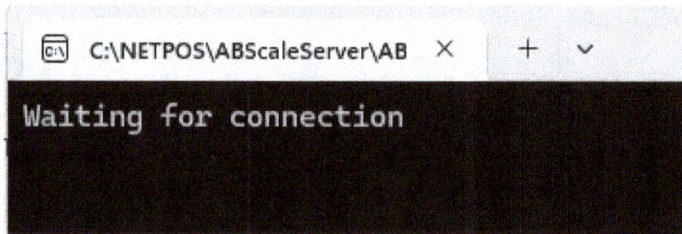

6. Open Visual Studio and open the EX82_Barcode_Scale project.
7. Start debugging the application. The barcode scanner should be found and claimed.
8. Scan some barcodes.
9. Put something on the scale, and click the "Get Weight" button. The weight will be retrieved and displayed in the console application and the GUI application.

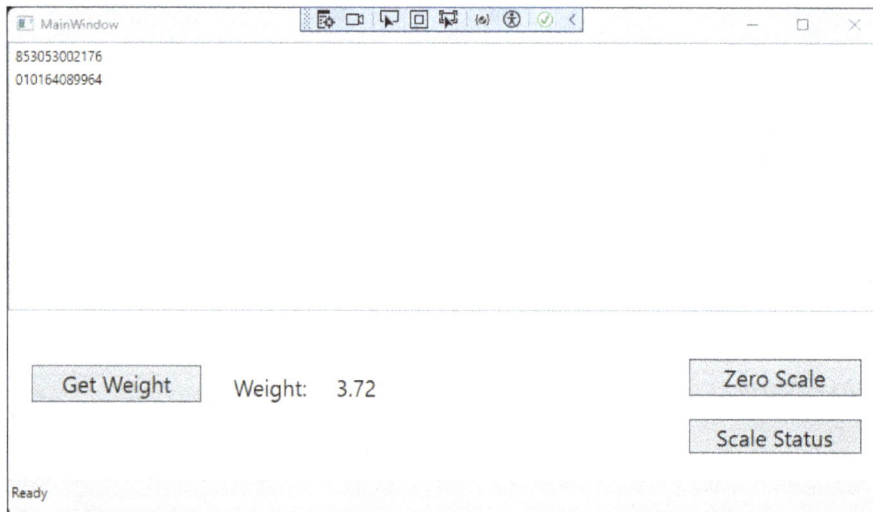

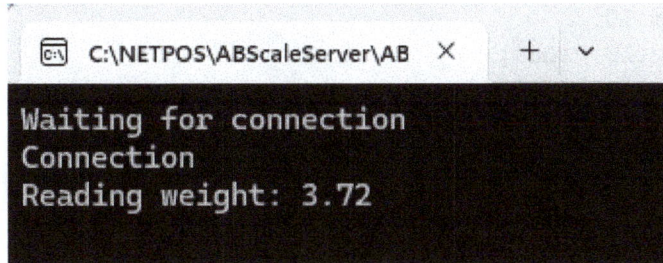

10. Click on the "Scale Status" button. An output of the status was not set up for the console, but the result will be displayed in the GUI.

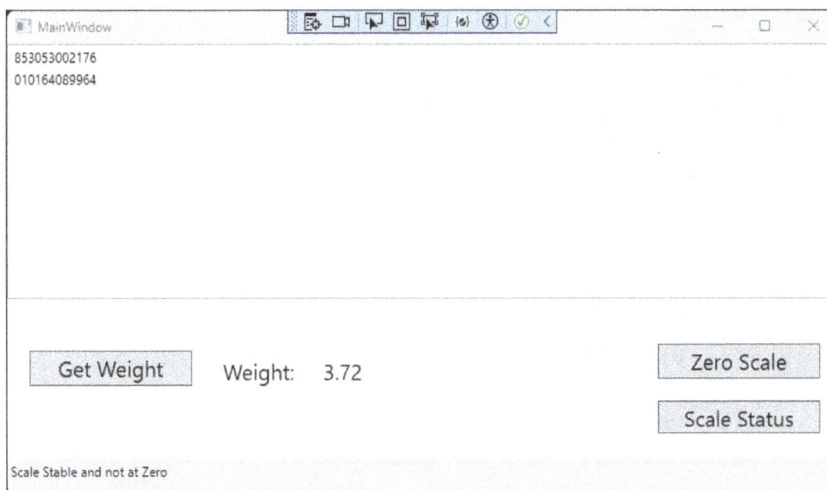

11. Stop both applications.

8.7 Exercise 8.4: Minor Additions

Running both applications separately is the best way to debug both applications. Once debugging is done, the next step is to roll them into a product where they work together.

8.7.1 Add Code to Hide the ABScaleServer

Seeing the console window is nice for debugging purposes, but in practice, the window should be hidden. In this section, we will add code to hide the console window.

1. In Visual Studio, with the ABScaleServer project open, open the Programs.cs file.
2. Add the following using statement for InteropServices:

```
using System.Threading.Tasks;
```

```
using System.IO;
using System.IO.Pipes;
using Microsoft.PointOfService;
using System.Runtime.InteropServices;
```

3. Now, we will add the code to hide the console window. The following code listing shows the additions to the program:

```
namespace ABScaleServer
{
    internal class Program
    {

        [DllImport("kernel32.dll")]
        static extern IntPtr GetConsoleWindow();

        [DllImport("user32.dll")]
        static extern bool ShowWindow(IntPtr hWnd, int nCmdShow);


        static private PosExplorer localPosExplorer;
        static private Scale localScale;

        static readonly int HIDECONSOLE = 0;
        static readonly int SHOWCONSOLE = 5;

        static async Task Main(string[] args)
        {
            var h = GetConsoleWindow();
            ShowWindow(h, HIDECONSOLE);
            localPosExplorer = new PosExplorer();
```

First, the DLL imports for kernel32 and user32.dll get access to the methods to be called. Next, two constants are created for HIDECONSOLE and SHOWCONSOLE. Just within the Main(), the handle of the console window is passed into the ShowWindows call along with the HIDECONSOLE value to hide the console window. If you ever need to debug this again, just swap out the HIDECONSOLE with SHOWCONSOLE.

4. Save the file.

8.7.2 Add code to EX82_Barcode_Scale Start and Stop ABScaleServer

Running both applications separately is good for testing. In this step will add code to start and stop ABScaleServer when EX82_Barcode_Scale is running:

1. Open Visual Studio.
2. Open the EX82_Barcode_Scale project.
3. Open MainWindows.xaml
4. Above the Grid tag, add Closed and Loaded events to the XAML includes. The event handlers will be generated in MainWindows.xaml.cs.

```xml
<Window x:Class="EX82_Barcode_Scale.MainWindow"

xmlns="http://schemas.microsoft.com/winfx/2006/xaml/presentatio
n"
        xmlns:x="http://schemas.microsoft.com/winfx/2006/xaml"

xmlns:d="http://schemas.microsoft.com/expression/blend/2008"
        xmlns:mc="http://schemas.openxmlformats.org/markup-
compatibility/2006"
        xmlns:local="clr-namespace:EX82_Barcode_Scale"
        mc:Ignorable="d"
        Title="MainWindow" Height="450" Width="800"
        Closed="Window_Closed"
        Loaded="Window_Loaded">
    <Grid>
```

5. OpenMainWindows.xaml.cs and fill in the code for the Window_Loaded event handler:

```csharp
private void Window_Loaded(object sender, RoutedEventArgs e)
{

Process.Start("C:\\NETPOS\\ABScaleServer\\ABScaleServer\\bin\\D
ebug\\ABScaleServer.exe");
}
```

The path entered in the code listing is an example. Please, change this to the correct path for your system. When creating a project, it is best to have both applications in the same folder so a get local path call can be used.

6. Add the code for the Window_Closed event handler:

```csharp
private async void Window_Closed(object sender, EventArgs e)
{
```

265

```
    try
    {
        var    pipeClient    =    new    NamedPipeClientStream(".",
"scalepipe", PipeDirection.InOut);
        pipeClient.Connect();

        var reader = new StreamReader(pipeClient);
        var writer = new StreamWriter(pipeClient);

        await writer.WriteLineAsync("close");
        await writer.FlushAsync();
        pipeClient.Close();
    }
    catch (Exception ex) {
        txtStatus.Text= ex.Message;
    }
}
```

7. Save the project.

8.7.3   Test the Changes

1. Start the debugger for the EX82_Barcode_Scale. The console window for ABScaleServer will briefly appear and get minimized to the taskbar.
2. Scan some barcodes and test the scale functionality. Everything should be working.
3. Close the EX82_Barcode_Scale window. This should close and remove the ABScaleServer console window from the taskbar. If you just stop debugging, the ABScaleServer console window will still be running and will have to be closed manually.

8.8   Is This the only solution?

The ABScaleServer project is a nice solution for testing and development, but creating a Windows Service would be better since there will be no console window to address. An event log would be used in place of the console output. Best of all the service can be controlled and monitored with SC.EXE or System.ServiceProcess in C#.

Since the Avery Berkel 6710 scale is a simple serial port device, there is no need to use POS for .NET at all. The application could use System.IO.Ports like the EX51_Scale_Test project to directly communicate with the scale. There is no need to monitor a separate application or service.

Finally, if the application has to interact with more than one of the remaining 31 POS devices, expanding the console services to accommodate other devices is possible.

8.9   POS WinRT versus POS for .NET – POS Device Selection

The goal of the book has been to explain the very basic fundamentals of accessing POS devices so you create a solution for your application. The importance of this chapter highlights a fundamental choice that has to be made when developing an application to access POS devices. POS WinRT offers a more modern approach to working with POS devices with .NET continuing to be advanced by Microsoft. POS for .NET requires .NET Framework, which will not be advanced. If there was a choice, the recommendation is to create applications with POS WinRT, and use a service or secondary application to support any POS devices with POS for .NET.

Keep in mind the choice of POS devices plays an important role. POS WinRT supports the 5 major POS devices, even so, you have to find the POS hardware that works with the built-in assemblies or POS devices that come with POS WinRT-supported assemblies. POS for .NET requires OPOS drivers or a Service Object. For devices like the Avery Berkel 6710 scale, a simple serial call can be made without the need for an OPOS driver or Service Object.

There are no simple answers because every project and every project's budget are different. Make sure to take the time and plan out the whole system accordingly including the product life-cycle.

8.10 Summary

POS WinRT allows for a more modern programming experience. POS WinRT supports the 5 major POS devices that most people use, but when it is time to support one or two of the remaining 31 devices, a return to POS for .NET and the .NET Framework is one possible solution. The chapter and exercises explored the idea of WPF .NET 8 applications interacting with the POS Scale device via a POS for .NET console application. Using Named Pipes, the two applications can communicate with each out. Care must be taken to address the situations when the server closes or is not available. This chapter closes out the major topics covering both POS for .NET and POS for WinRT. The final chapter will explore putting all the devices together in a couple of projects.

9 A Few Ideas

The setup and access POS device details using POS WinRT and POS for .NET have been covered in the previous chapters. This chapter presents a few ideas and more complete examples. The chapter will cover the following:

- Sharing a network Receipt Printer between two computers.
- Demonstrate how to save a paired device path to be re-connected on application restart for a POS WinRT .NET application.
- Inventory and Cash Register applications.

9.1 Exercise 9.1: Share Receipt Printer and Save/Restore POS Device

Receipt printers are expensive POS devices. Having the ability to share a receipt printer between two or more systems has been a popular topic for some time. In this exercise, an application will be created that can run on multiple systems and connect to a shared network printer. The snapshot enumeration method will be used to search and find network receipt printers. The selected printer path will be saved to a registry key to be re-used when the application runs again.

9.1.1 Create the Visual Studio Project

The first step is to create the project and perform the basic setup tasks:

1. Open Visual Studio.
2. Create a new project.
3. Search for and select "WPF Application for C#" and click Next.
4. Name the project EX91_PrinterShared and click Next.
5. Keep the .NET version at ".NET 8.0 (Long Term Support)" and click Create.
6. The next and most important step: from the menu select Project-> EX91_PrinterShared properties.

7. Under General, change the following:

Target OS Version: 10.0.26100.0
Supported OS Version: 10.0.19041.0

The reason the OS version is important is that the minimum OS version of 10.0.19041.0 is the version that supports the POS WinRT namespaces. If you use any OS version below the minimum, you will not be able to get access to the POS WinRT namespaces.

8. Save the project when finished.
9. Close the EX91_PrinterShared properties page.

9.1.2   Set Up the XAML Controls

Now we will set up the UI.

1. In Solution Explorer, open the MainPage.xaml.
2. Add a TextBox control.
    b.   Name: txtMsg
3. Add a Button to the right side of the TextBox.
    c.   Name: btnPrint
    d.   Content: Print Message
4. Add a ListBox below the two previous controls.
    a.   Name: lstPrinters
5. Add a Button below the ListBox.
    a.   Name: PrinterRefresh
    b.   Content: Printer Refresh
6. Add the Status strip, and resize it to fit the bottom of the page.
    a.   Name: sBar
7. Add a TextBlock control to the status strip.
    d.   Name: txtStatus
    e.   Text: Ready
    f.   Font: 10px

Here is the XAML code:

```xml
<Window x:Class="EX91_PrinterShared.MainWindow"

xmlns="http://schemas.microsoft.com/winfx/2006/xaml/presentatio
n"
        xmlns:x="http://schemas.microsoft.com/winfx/2006/xaml"

xmlns:d="http://schemas.microsoft.com/expression/blend/2008"
        xmlns:mc="http://schemas.openxmlformats.org/markup-
compatibility/2006"
```

270

```
        xmlns:local="clr-namespace:EX91_PrinterShared"
        mc:Ignorable="d"
        Title="MainWindow" Height="450" Width="800">
    <Grid>
        <ListBox                                x:Name="lstPrinters"
d:ItemsSource="{d:SampleData  ItemCount=5}"  Margin="0,267,0,91"
SelectionChanged="lstPrinters_SelectionChanged"/>
        <Button          x:Name="btnPrint"              Content="Print"
HorizontalAlignment="Left"                    Margin="629,48,0,0"
VerticalAlignment="Top"  Width="124"  Height="84"  FontSize="16"
Click="btnPrint_Click"/>
        <TextBox     x:Name="txtMsg"     HorizontalAlignment="Left"
Margin="73,72,0,0" TextWrapping="Wrap" Text="Enter a message"
VerticalAlignment="Top" Width="497" Height="36"/>
        <StatusBar x:Name="sBar" Margin="0,387,0,0">
            <TextBlock   x:Name="txtStatus"    TextWrapping="Wrap"
Text="Ready" Width="792" Height="34"/>
        </StatusBar>

    </Grid>
</Window>
```

The user will simply enter some text in the upper textbox and push the button to send the data to the printer. The list box will present the available network Receipt Printers to connect to. The Designer should show the following:

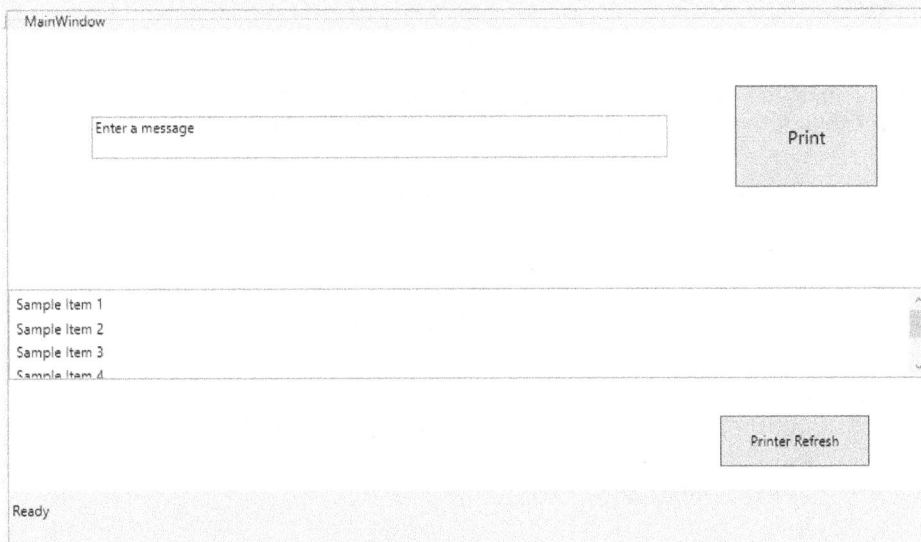

8. Save the MainWindow.xaml file.

9.1.3 Write the code

Now we will create the code behind the interface.

1. Open MainWindow.xaml.cs file.
2. Add the following using statements:

```
using Windows.Devices.PointOfService;
using Windows.Devices.Enumeration;
using System.Collections;
using Microsoft.Win32;
```

3. Within the MainWindow Class and before the MainWindow() method, add the following global variables:

```
public partial class MainWindow : Window
{

    PosPrinter myPrinter = null;
    ClaimedPosPrinter myClaimedPrinter = null;
    ArrayList myDeviceIDs = new ArrayList();
    static string savedPrinter = "";


    public MainWindow()
```

The printer and claimed POS printer are the same as in the previous exercises. The ArrayList is going to hold the POS Printer device path while the ListBox will show the name. The SavePrinter string will be used to store the path for the selected printer.

4. Below the MainWindows() method, create the POSSetup() method:

```
private async void POSSetup()
{
    txtStatus.Text = "Searching for printers. Please wait...";

    RegistryKey                posPrinterPath                =
Registry.CurrentUser.OpenSubKey(@"SOFTWARE\PosPrinterShared")!;

    if (posPrinterPath != null) {
```

272

```
        savedPrinter                                      =
posPrinterPath.GetValue("posPrinterPath").ToString();
        txtStatus.Text  =  "Previous  printer  ID  found  "  +
savedPrinter;
    }
    else
    {
        posPrinterPath                                    =
Registry.CurrentUser.CreateSubKey(@"SOFTWARE\PosPrinterShared")
;

        DeviceInformationCollection  posPrinters  =  await
DeviceInformation.FindAllAsync(PosPrinter.GetDeviceSelector(Pos
ConnectionTypes.IP));

        foreach (DeviceInformation deviceinfo in posPrinters)
        {
            lstPrinters.Items.Add(deviceinfo.Name);
            myDeviceIDs.Add(deviceinfo.Id);
        }
    }
}
```

The first thing that happens is to check for any previously connected printers. If the registry subkey exists, the value is retrieved. If the registry subkey doesn't exist, the subkey will be created and the search for all network IP POS printers will be performed. The snapshot method is faster when filtering for only network IP printers. The ListBox is filled in with all the POS printers found. A claim for the printer is not made here.

5. Go back to MainWindow.xaml Designer view.
6. Double-click on the ListBox control to generate the event handler.
7. Double-click on both buttons to generate the button click event handlers.
8. Fill in the code for the lstPrinters_SelectionChanged event handler:

```
private async void lstPrinters_SelectionChanged(object sender,
SelectionChangedEventArgs e)
{

    ListBox? lb = sender as ListBox;
    if (lb != null) {
        txtStatus.Text = lb.SelectedItem.ToString();
        myPrinter                       =               await
PosPrinter.FromIdAsync((string)myDeviceIDs[lb.SelectedIndex]);
```

273

```
        if (myPrinter != null) {
            txtStatus.Text    =    "Printer    sected:    "    +
myPrinter.DeviceId;
            myClaimedPrinter                         =                await
myPrinter.ClaimPrinterAsync();

            if (myClaimedPrinter != null) {

                savedPrinter                                        =
myDeviceIDs[lb.SelectedIndex].ToString();        //Saving        he
deviceinfo.Id
                RegistryKey        posPrinterPath        =
Registry.CurrentUser.OpenSubKey(@"SOFTWARE\PosPrinterShared",
true)!;
                posPrinterPath.SetValue("posPrinterPath",
savedPrinter.ToString());

                txtStatus.Text = "Printer saved. Ready for use.";
                myClaimedPrinter.Dispose();
                myPrinter.Dispose();
            }
            else
            {
                txtStatus.Text = "Printer Claim test failed! Try
again";
            }
        }
        else
        {
            txtStatus.Text = "Printer not found";
        }
    }
    else
    {
        txtStatus.Text = "List error";
    }
}
```

When the user clicks on a printer in the list box, there is an attempt to claim the printer. If the claim is successful, the device path is saved locally and stored in the registry subkey. Immediately, the claim to the printer will be released since we want other systems to get access to the printer, but the paring will remain. Any failures to claim or find the printer will be displayed in the status box.

9. Next, add the code for the btnPrint_Click event handler:

```
private async void btnPrint_Click(object sender, RoutedEventArgs e)
{

    if (savedPrinter != null)
    {
        myPrinter                           =                  await
PosPrinter.FromIdAsync(savedPrinter.ToString());

    }
    else
    {
        txtStatus.Text = "Not connected to a printer";
    }

    if (myPrinter != null) {

        txtStatus.Text = "Printer Found " + myPrinter.DeviceId;
        myClaimedPrinter = await myPrinter .ClaimPrinterAsync();
        if (myClaimedPrinter != null) {

            await myClaimedPrinter.RetainDeviceAsync();

            if (await myClaimedPrinter.EnableAsync()) {

                txtStatus.Text  =  "Printer  is  claimed  and
enabled";

                ReceiptPrintJob              job              =
myClaimedPrinter.Receipt.CreateJob();
                job.PrintLine(txtMsg.Text);
                job.PrintLine("");
                job.PrintLine("");
                job.PrintLine("");
                job.PrintLine("");
                job.PrintLine("");
                job.PrintLine("");
```

```
                job.CutPaper();
                if (!await job.ExecuteAsync())
                {
                    txtStatus.Text = "Print failed";
                }

                myClaimedPrinter.Dispose();
                myPrinter.Dispose();

                txtStatus.Text = " Print job completed. Printer
released";

            }
            else
            {
                txtStatus.Text = " Printer is NOT claimed and
enabled";
            }
        }
        else
        {
            txtStatus.Text = "Claim Printer failed";
        }
    }
    else
    {
        txtStatus.Text = "Printer Not Found";
    }
}
```

The secret to sharing POS Receipt Printer devices is to perform the quick operation to claim, print, and close the claim all at once. The quick claim-print-and-release method allows other systems to connect and print to the Receipt Printer. Any failures will show up in the status box. Besides performing the claim on the printer, the print operation is the same as Exercise 7.3.

As quick as the operation is, there is a possibility that two or more systems could attempt to access the printer at the exact same time. Code can be added to wait in a loop for the printer to be freed up and the operation retried.

10. Add the code to the btnPrinterRefresh_Click event handler to refresh the POS printer ListBox:

```
private async void btnPrinterRefresh_Click(object sender,
RoutedEventArgs e)
{
    DeviceInformationCollection posPrinters = await
DeviceInformation.FindAllAsync(PosPrinter.GetDeviceSelector(Pos
ConnectionTypes.IP));

lstPrinters.Items.Clear();

    foreach (DeviceInformation deviceinfo in posPrinters)
    {

        lstPrinters.Items.Add(deviceinfo.Name);
        myDeviceIDs.Add(deviceinfo.Id);
    }
}
```

11. Finally add the call to POSSetup() in mainWindows():

```
public MainWindow()
{
    InitializeComponent();
    POSSetup();
}
```

12. Save the project.

9.1.4 Test the Application on Multiple Systems
At least two systems connected to the same network are required.

1. Make sure the POS Receipt Printer is connected to the network. Go back to Exercise 7.3 if necessary.
2. Build the application and correct any errors.
3. Copy the application to a different system.
4. Optional: you may have to install .NET 8 runtime and VC 2022 runtime for the application to run on this other system.
5. Start the application on the development machine in debug mode, and start the application on the other system.
6. The printer list will fill in quickly since the search is for IP POS Printers. Select the same printer on both systems.
7. A message box will appear to pair the POS printer with the system. Select OK.

277

8. Type a message in one system and click the Print button. The Receipt Printer should print the message.

9. Do the same for the other system. The Receipt Printer should print the message.

10. For one of the systems, close the application and run the application again. This time the application uses the stored path in the registry and the printer list is not populated. A pairing dialog doesn't appear since the connection is with the saved path.

11. Type a message in one system and click the Print button. The Receipt Printer should print the message.

12. Open the registry editor and go to HKCU\Software\PosPrinterShared to see the value store in registry.

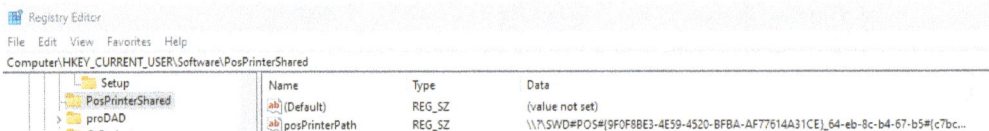

13. Delete the registry subkey since it will not be needed any more.

14. Close registry editor and the applications when finished.

The exercise demonstrated a different approach to the snapshot enumeration method by filtering for specific connected devices. Also, the ability to store the select POS printer avoids having to pair the printer each time the application starts. The printer list box is still there to select a different printer if required.

9.2 Store Solution: Inventory and Cash Register Applications

The next three exercises will create a store inventory and cash register solution. Both applications will use the same SQLite database. The inventory application will be used to update the database with items in stock. The cash register application will handle sales. For these exercises, both applications exist on the same computer, but ideally, they would run on separate computers. These exercises will produce a "real-enough" solution, but they are not a fully implemented solution. There is room for improvement. SQLite is only being used for demonstration. Using SQL Express or a database in the cloud would be more efficient and robust.

9.3 Exercise 9.2: Database Setup

SQLite will be the database engine for the solution. There is SQLite Browser available on GitHub: https://github.com/sqlitebrowser/sqlitebrowser that we will use to create the database.

1. Download and install the SQLite Browser from https://github.com/sqlitebrowser/sqlitebrowser. Scroll down the information pages to download the latest pre-built application.
2. After installation is complete, launch the DB Browser (SQLite) application.
3. Click on "New Database" from the toolbar.
4. You will be asked to create a name and location for the database file.
    a. Name: Inventory
    b. Location: C:\NETPOS
5. Fill in the Table name: Inventory
6. Next, you need to add all the fields for the Inventory table, click Add:
    a. Name: Item, Type: Integer, PK box checked
    b. Name: ProductName, Type: Text
    c. Name: Barcode, Type: Text
    d. Name: Price, Type: Numeric
    e. Name: Qty, Type: Integer
7. Click OK when finished.

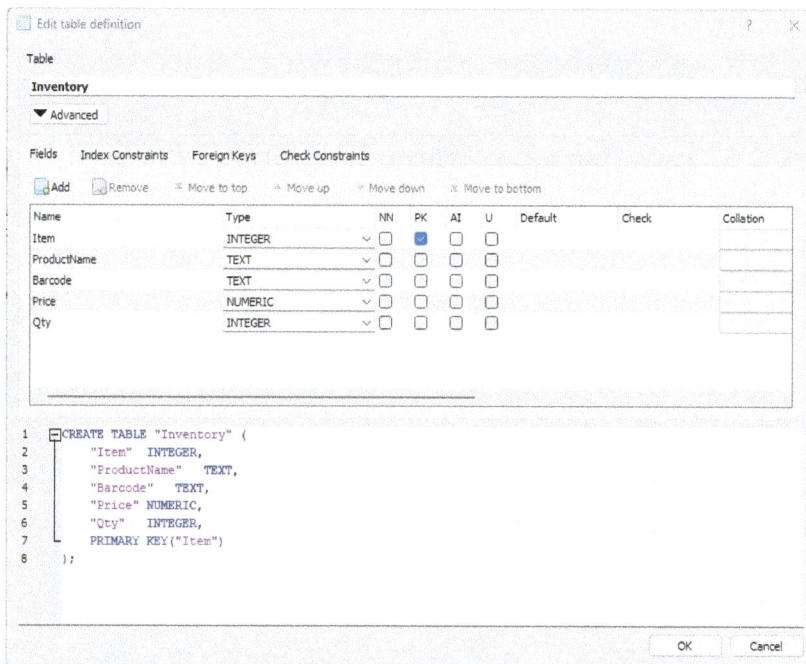

8. Close DB Browser (SQLite) when finished. If asked to save the database, click Yes.

9.4   Exercise 9.3: InvScanner

With the database setup, the first application will be an inventory scanner. The only POS device for this application is a barcode scanner, which will be used to lookup an item in the database, add a new item, or increase/decrease the quantity of an item in inventory.

9.4.1   Create the Visual Studio Project

The application will be a WPF .NET 8 application. The first step is to create the project and perform the basic setup tasks. In addition, the NuGet package for SQLite will be added.

1. Open Visual Studio.
2. Create a new project.
3. Search for and select "WPF Application for C#" and click Next.
4. Name the project InvScanner and click Next.
5. Keep the .NET version at ".NET 8.0 (Long Term Support)" and click Create.
6. The next and most important step: from the menu select Project-> InvScanner properties.
7. Under General, change the following:

> Target OS Version: 10.0.26100.0
> Supported OS Version: 10.0.19041.0

The reason the OS version is important is that the minimum OS version of 10.0.19041.0 is the version that supports the POS WinRT namespaces. If you use any OS version below the minimum, you will not be able to get access to the POS WinRT namespaces.

8. Close the InvScanner properties page.
9. The SQLite NuGet package needs to be pulled in from the Internet. From the menu select Tools -> NuGet package Manager-> Manage NuGet Packages for Solution.
10. A new NuGet – Solution page appears, click Browse.
11. In the search box enter "Microsoft.Data.SQLite".
12. Scroll down to the "Microsoft.Data.SQLite".
13. Install the latest stable release.
14. Click OK to make the changes, and click to accept the license.
15. Save the whole project.
16. Close the NuGet – Solution page.

9.4.2   Set Up the XAML Controls

Now, we will set up the UI.

1. In Solution Explorer, open the MainWindow.xaml.
2. Add a Label in the top, left corner.
   a.   Name: lblName

       b.   Content: Product Name:

3. Add a Textbox next to the label.
   a. Name: txtProductName
   b. Text: <empty>

4. Below both the label and text box, add a Label
   a. Name: lblBarcode
   b. Content: Barcode:

5. Add a Textbox next to the label
   a. Name: txtBarcode
   b. Text: <empty>

6. Below both the label and text box, add a Label
   a. Name: lblPrice
   b. Content: Price:

7. Add a Textbox next to the label
   a. Name: txtPrice
   b. Text:<empty>

8. Below both the label and text box, add a Label
   a. Name: lblQty
   b. Content: Qty:

9. Add a Textbox next to the label
   a. Name: txtQty
   b. Text:<empty>

10. Add a radio Radio Button to the left of the text boxes.
   a. Name: rbListItem
   b. Content: List Item
   c. IsChecked: Checked

11. Add a radio Radio Button below the first radio button
   a. Name: rbNew
   b. Content: New Item

12. Add a radio Radio Button below the second radio button
   a. Name: rbIncrement
   b. Content: Increment QTY

13. Add a radio Radio Button below the third radio button
   a. Name: rbDecrement
   b. Content: Decrement QTY

14. Add a Data grid under the other controls and leave room for the status bar.
   a. Name: dgInventory
   b. In the XAML code, add AutoGenerateColumns="False"

15. Add a StatusBar at the bottom of the page.
   a. Name: sBar

16. Add a TextBlock to the StatusBar.
   a. Name: txtStatus

Here is the XAML code:

```
<Window x:Class="InvScanner.MainWindow"

xmlns="http://schemas.microsoft.com/winfx/2006/xaml/presentatio
n"
        xmlns:x="http://schemas.microsoft.com/winfx/2006/xaml"

xmlns:d="http://schemas.microsoft.com/expression/blend/2008"
        xmlns:mc="http://schemas.openxmlformats.org/markup-
compatibility/2006"
        xmlns:local="clr-namespace:InvScanner"
        mc:Ignorable="d"
        Title="MainWindow" Height="450" Width="800">
    <Grid>
        <DataGrid                        x:Name="dgInventory"
AutoGenerateColumns="False"        d:ItemsSource="{d:SampleData
ItemCount=5}" Margin="0,156,0,54"/>
        <Label     x:Name="lblName"     Content="Product     Name:"
HorizontalAlignment="Left"                    Margin="10,10,0,0"
VerticalAlignment="Top" Width="135"/>
        <TextBox                        x:Name="txtProductName"
HorizontalAlignment="Left"                    Margin="122,15,0,0"
TextWrapping="Wrap" VerticalAlignment="Top" Width="395"/>
        <Label      x:Name="lblBarcode"     Content="Barcode:"
HorizontalAlignment="Left"                    Margin="35,45,0,0"
VerticalAlignment="Top"/>
        <TextBox x:Name="txtBarcode" HorizontalAlignment="Left"
Margin="122,50,0,0" TextWrapping="Wrap" VerticalAlignment="Top"
Width="395"/>
        <Label           x:Name="lblPrice"         Content="Price:"
HorizontalAlignment="Left"                    Margin="52,81,0,0"
VerticalAlignment="Top"/>
        <TextBox    x:Name="txtPrice"    HorizontalAlignment="Left"
Margin="122,85,0,0" TextWrapping="Wrap" VerticalAlignment="Top"
Width="198"/>
        <Label             x:Name="lblQty"         Content="Qty:"
HorizontalAlignment="Left"                    Margin="55,116,0,0"
VerticalAlignment="Top"/>
        <TextBox     x:Name="txtQty"     HorizontalAlignment="Left"
Margin="122,120,0,0" TextWrapping="Wrap" VerticalAlignment="Top"
Width="198"/>
        <StatusBar Margin="0,385,0,0">
```

```
            <TextBlock    x:Name="txtStatus"    TextWrapping="Wrap"
Text="Ready" Width="775" Height="36"/>
        </StatusBar>
        <RadioButton      x:Name="rbNew"      Content="New      Item"
HorizontalAlignment="Left"                 Margin="665,62,0,0"
VerticalAlignment="Top"/>
        <RadioButton    x:Name="rbIncrement"    Content="Increment
QTY"        HorizontalAlignment="Left"      Margin="665,89,0,0"
VerticalAlignment="Top"/>
        <RadioButton    x:Name="rbDecrement"    Content="Decrement
QTY"        HorizontalAlignment="Left"     Margin="665,117,0,0"
VerticalAlignment="Top"/>
        <RadioButton    x:Name="rbListItem"    Content="List    Item"
HorizontalAlignment="Left"                 Margin="665,33,0,0"
VerticalAlignment="Top" IsChecked="True"/>

    </Grid>
</Window>
```

Here is what the Designer should look like:

17. Save the MainWindow.xaml file.

9.4.3   Write the code
Now we will put the code behind the interface.

283

1. Open the MainWindow.xaml.cs file.
2. Add the following using statements so we can access the PointOfService namespace, the namespaces to enumerate the POS devices, and access the SQLite APIs:

```
using System.Windows;
using System.Windows.Controls;
using System.Windows.Data;
using Microsoft.Data.Sqlite;
using System.Diagnostics;
using Windows.Devices.PointOfService;
using Windows.Devices.Enumeration;
using Windows.Storage.Streams;
using System.Data;
using System.Collections.ObjectModel;
```

3. Above the MainWindow() method, add the global properties for a BarcodeScanner, ClaimedBarcodeScanner, DevcieWatcher, and the DataTable.

```
DataTable inventoryTable = new DataTable();
BarcodeScanner scanner = null;
ClaimedBarcodeScanner claimedScanner = null;
DeviceWatcher posDevWatcher =
DeviceInformation.CreateWatcher(BarcodeScanner.GetDeviceSelecto
r(PosConnectionTypes.All));
Int64 maxIndex = 0;
```

4. After the MainPage() method, create the POSSetup() method and add the following code to create the Device Watcher events:

```
private void POSSetup()
{
    posDevWatcher.Added += PosDevWatcher_Added;
    posDevWatcher.Removed += PosDevWatcher_Removed;
    posDevWatcher.Start();
}
```

Remember, when you add the Added and Removed events, after the += hit the Tab key twice to generate the events. The events are listed before the watcher is started. Once the watcher starts, it will continue to monitor for devices being added and removed from the system until either you stop the watcher or the application session is ended.

Before we fill in the code for the event handlers, two methods must be added to address the DataGrid control. The first is called FillDataGrid(), which will be called when the application first starts up, The second is the DataGridRefresh(), which will be called when there is a change of data to the database.

5. Create the new FillDataGrid() method MainWindow() and before POSSetup(), fill in the following code:

```
private void FillDataGrid()
{

    inventoryTable.Columns.Add("Index", typeof(string));
    inventoryTable.Columns.Add("ProductName", typeof(string));
    inventoryTable.Columns.Add("Barcode", typeof(string));
    inventoryTable.Columns.Add("Price", typeof(string));
    inventoryTable.Columns.Add("Qty", typeof(string));

    //Add the columns
    for (int i = 0; i < inventoryTable.Columns.Count; i++)
    {
        dgInventory.Columns.Add(new DataGridTextColumn()
        {
            Header = inventoryTable.Columns[i].ColumnName,
            Binding = new Binding { Path = new PropertyPath("["
+ i.ToString() + "]") }
        });
    }

    Int64 recordIndex = 1;

    string dbpath = "c:\\NETPOS\\Inventory.db";
    try
    {
        using       (SqliteConnection       db       =       new
SqliteConnection($"Filename={dbpath}"))
        {
            db.Open();

            if (db.State == System.Data.ConnectionState.Open)
            {
```

```
                using        (SqliteCommand      cmd     =       new
SqliteCommand("SELECT COUNT(Item)  FROM Inventory", db))
                {
                    maxIndex = (Int64)cmd.ExecuteScalar();

                    while (recordIndex <= maxIndex)
                    {

                    using    (SqliteCommand   cmd2   =   new
SqliteCommand("SELECT*  FROM  Inventory  WHERE  Item  =  "  +
recordIndex.ToString(), db))
                        {
                            using (SqliteDataReader   reader   =
cmd2.ExecuteReader())
                            {
                                reader.Read();

inventoryTable.Rows.Add(reader["Item"].ToString(),
reader["ProductName"].ToString(),  reader["Barcode"].ToString(),
reader["Price"].ToString(), reader["Qty"].ToString());
                            }
                        }
                        recordIndex++;
                    }
                }
            }
            db.Close();
            dgInventory.ItemsSource                              =
inventoryTable.DefaultView;

        }
    }
    catch (Exception ex)
    {
        Debug.WriteLine("Exception: " + ex.ToString());
    }
}
```

The purpose of the DataGrid is to provide a view of all the items in the database. The code creates the DataTable and imports all the data from the SQLite database. The final step is to set the DataTable as the source for the DataGrid. Ideally, it would have been nice to have an SQLIteAdapter do a more direct link, but SQL Adapter support is missing from the

NuGet package. The table acts as the go-between, which is not ideal. Other SQL engines have more robust support than SQLite.

6. Below the FillDataGrid() method, create the DataGridRefresh() method and fill in the following code:

```
private void DataGridRefresh()
{
    inventoryTable.Rows.Clear();
    Int64 recordIndex = 1;
    string dbpath = "c:\\NETPOS\\Inventory.db";
    try
    {
        using        (SqliteConnection        db        =        new
SqliteConnection($"Filename={dbpath}"))
        {
            db.Open();

            if (db.State == System.Data.ConnectionState.Open)
            {

                using        (SqliteCommand        cmd        =        new
SqliteCommand("SELECT COUNT(Item)  FROM Inventory", db))
                {
                    maxIndex = (Int64)cmd.ExecuteScalar();

                    while (recordIndex <= maxIndex)
                    {

                    using    (SqliteCommand    cmd2    =    new
SqliteCommand("SELECT*    FROM    Inventory    WHERE    Item    =    "    +
recordIndex.ToString(), db))
                        {
                        using    (SqliteDataReader    reader    =
cmd2.ExecuteReader())
                            {
                                reader.Read();

inventoryTable.Rows.Add(reader["Item"].ToString(),
reader["ProductName"].ToString(),  reader["Barcode"].ToString(),
reader["Price"].ToString(), reader["Qty"].ToString());
                            }
```

```
                               }
                          recordIndex++;
                      }
                  }
              }
              db.Close();

              dgInventory.Items.Refresh();

          }
      }
      catch (Exception ex)
      {
          Debug.WriteLine("Exception: " + ex.ToString());
      }
  }
}
```

When there is a change to the database, such as a new item or an increment or decrement of quantity, the method is called to refresh the DataGrid with the new data. The DataTable is cleared and all the content in the database is filled into the DataTable. Again, this is not the most efficient solution, especially for large datasets.

7. For the PosDevWatcher_Added event handler add the following:

```
private async void PosDevWatcher_Added(DeviceWatcher sender,
DeviceInformation args)
{
    scanner = await BarcodeScanner.FromIdAsync(args.Id);
    if (scanner != null)
    {

        claimedScanner = await scanner.ClaimScannerAsync();
        if (claimedScanner != null)
        {
            scanner.StatusUpdated += Scanner_StatusUpdated;
            claimedScanner.ReleaseDeviceRequested            +=
ClaimedScanner_ReleaseDeviceRequested;
            claimedScanner.DataReceived                      +=
ClaimedScanner_DataReceived;
            claimedScanner.IsDecodeDataEnabled = true;
            await claimedScanner.EnableAsync();
```

```
            Application.Current.Dispatcher.Invoke((Action)(()
=> txtStatus.Text = "Barcode Scanner Claimed: " + args.Id));


        }
        else
        {
            Application.Current.Dispatcher.Invoke((Action)(()
=> txtStatus.Text = "Barcode Scanner Not Claimed"));
        }

    }
    else
    {
        Application.Current.Dispatcher.Invoke((Action)(()       =>
txtStatus.Text = "Barcode Scanner Not Found"));
    }
}
```

The Device Watcher added event checks for a barcode scanner is connected, and then performs all the same steps to enumerate the device as before.

8. For the PosDevWatcher_Remove event handler add the following:

```
private     void     PosDevWatcher_Removed(DeviceWatcher     sender,
DeviceInformationUpdate args)
{
    claimedScanner.DataReceived -= ClaimedScanner_DataReceived;
    claimedScanner.ReleaseDeviceRequested                      -=
ClaimedScanner_ReleaseDeviceRequested;
    scanner.StatusUpdated -= Scanner_StatusUpdated;
    claimedScanner.Dispose();
    scanner.Dispose();
    Application.Current.Dispatcher.Invoke((Action)(()          =>
txtStatus.Text = "Barcode Scanner Removed"));
}
```

The handler actually destroys the claimedScaner and scanner objects by calling the Dispose() method. The Dispose() method could have been used in the previous exercises, but a newly attached device would require a manual operation to connect to the device. Device Watcher will automatically connect to the newly attached barcode scanner.

13. Add the code for the ClaimedScanner_DataReceived event handler method:

```
private void ClaimedScanner_DataReceived(ClaimedBarcodeScanner
sender, BarcodeScannerDataReceivedEventArgs args)
{
    //Read the data from the buffer and convert to a string.
    var                    scanDataLabelReader                 =
DataReader.FromBuffer(args.Report.ScanDataLabel);
    string                 stringDataLabel                     =
scanDataLabelReader.ReadString(args.Report.ScanDataLabel.Length
);

    //Remove UPCe if capture by scanner
    if (stringDataLabel.Length > 13)
    {
        stringDataLabel = stringDataLabel.Substring(0, 13);
    }

    Int64 recordIndex = 1;

    string dbpath = "c:\\NETPOS\\Inventory.db";

    try
    {
        using      (SqliteConnection      db       =       new
SqliteConnection($"Filename={dbpath}"))
        {

            db.Open();

                if(db.State == System.Data.ConnectionState.Open)
                {

                    if
(Dispatcher.Invoke(()=>rbListItem.IsChecked.Equals(true)))
                    {
                        using   (SqliteCommand   cmd   =   new
SqliteCommand("SELECT* FROM Inventory WHERE Barcode = " +
stringDataLabel, db))
                        {
                            using   (SqliteDataReader   reader  =
cmd.ExecuteReader())
                            {
                                reader.Read();
```

```
                                recordIndex                =
Int64.Parse(reader["Item"].ToString());

                                string          pn          =
reader["ProductName"].ToString();
                                string          bc          =
reader["Barcode"].ToString();
                                string          price       =
reader["Price"].ToString();
                                string          qty         =
reader["Qty"].ToString();

                                Dispatcher.BeginInvoke((Action)(()
=> txtProductName.Text = pn));
                                Dispatcher.BeginInvoke((Action)(()
=> txtBarcode.Text = bc));
                                Dispatcher.BeginInvoke((Action)(()
=> txtPrice.Text = price));
                                Dispatcher.BeginInvoke((Action)(()
=> txtQty.Text = qty));

                        }
                    }
                db.Close();
            }

                if          (Dispatcher.Invoke(()         =>
rbNew.IsChecked.Equals(true)))
                    {


if(String.IsNullOrEmpty(Dispatcher.Invoke(() => txtPrice.Text))
|| String.IsNullOrEmpty(Dispatcher.Invoke(() => txtQty.Text)) ||
String.IsNullOrEmpty(Dispatcher.Invoke(()                     =>
txtProductName.Text)))
                        {
                            Dispatcher.BeginInvoke((Action)(()   =>
txtStatus.Text = "Values cannot be null"));

                        }
                        else
                        {
```

291

```
                    SqliteCommand      addData      =     new
SqliteCommand("INSERT    INTO    Inventory(ProductName,   Barcode,
Price, Qty)  Values(@pn1, @bc1, @pr1, @qty1)", db);

                    addData.Parameters.AddWithValue("@pn1",
Dispatcher.Invoke(() => txtProductName.Text.ToString()));
                    addData.Parameters.AddWithValue("@bc1",
stringDataLabel);
                    addData.Parameters.AddWithValue("@pr1",
float.Parse(Dispatcher.Invoke(()                            =>
txtPrice.Text.ToString())).ToString("0.00"));

addData.Parameters.AddWithValue("@qty1",
int.Parse(Dispatcher.Invoke(() => txtQty.Text.ToString())));

                    addData.ExecuteReader();

                    db.Close();

                    Dispatcher.BeginInvoke((Action)(()      =>
DataGridRefresh()));

                    Dispatcher.BeginInvoke((Action)(()      =>
txtStatus.Text = "Item added"));

                }

            }


            if          (Dispatcher.Invoke(()              =>
rbIncrement.IsChecked.Equals(true)))
                {
                Int32 qtyValue = 0;


                using   (SqliteCommand   cmd   =   new
SqliteCommand("SELECT*   FROM   Inventory   WHERE   Barcode  =  "  +
stringDataLabel, db))
                {
```

```
                    using     (SqliteDataReader    reader    =
cmd.ExecuteReader())
                    {
                        reader.Read();
                        recordIndex                          =
Int64.Parse(reader["Item"].ToString());

                        string           pn           =
reader["ProductName"].ToString();
                        string           bc           =
reader["Barcode"].ToString();
                        string           price        =
reader["Price"].ToString();
                        string           qty          =
reader["Qty"].ToString();

                        qtyValue = Int32.Parse(qty);

                        Dispatcher.BeginInvoke((Action)(()
=> txtProductName.Text = pn));
                        Dispatcher.BeginInvoke((Action)(()
=> txtBarcode.Text = bc));
                        Dispatcher.BeginInvoke((Action)(()
=> txtPrice.Text = price));
                        Dispatcher.BeginInvoke((Action)(()
=> txtQty.Text = qty));
                    }
                }

                Thread.Sleep(2000);
                using     (SqliteCommand    cmd    =    new
SqliteCommand("UPDATE Inventory Set Qty=@qty WHERE BarCode = " +
stringDataLabel, db))
                {
                    Int32 qtyIncrement = qtyValue + 1;
                    cmd.Parameters.AddWithValue("@qty",
qtyIncrement);
                    Dispatcher.BeginInvoke((Action)(()    =>
txtQty.Text = qtyIncrement.ToString()));
                    cmd.ExecuteNonQuery();
                }
            db.Close();
```

293

```
                    Dispatcher.BeginInvoke((Action)(()          =>
DataGridRefresh()));

                    Dispatcher.BeginInvoke((Action)(()          =>
txtStatus.Text = "Qty increased"));

            }

            if(Dispatcher.Invoke(()                             =>
rbDecrement.IsChecked.Equals(true)))
                {
                    Int32 qtyValue = 0;

                    using   (SqliteCommand   cmd   =    new
SqliteCommand("SELECT*  FROM  Inventory  WHERE  Barcode  =  "  +
stringDataLabel, db))
                    {
                        using    (SqliteDataReader   reader   =
cmd.ExecuteReader())
                        {
                            reader.Read();
                            recordIndex                        =
Int64.Parse(reader["Item"].ToString());

                            string          pn                 =
reader["ProductName"].ToString();
                            string          bc                 =
reader["Barcode"].ToString();
                            string         price               =
reader["Price"].ToString();
                            string          qty                =
reader["Qty"].ToString();
                            qtyValue = Int32.Parse(qty);

                            Dispatcher.BeginInvoke((Action)(()
=> txtProductName.Text = pn));
                            Dispatcher.BeginInvoke((Action)(()
=> txtBarcode.Text = bc));
                            Dispatcher.BeginInvoke((Action)(()
=> txtPrice.Text = price));
```

```
                              Dispatcher.BeginInvoke((Action)(()
=> txtQty.Text = qty));
                    }
            }

            using    (SqliteCommand    cmd    =    new
SqliteCommand("UPDATE Inventory Set Qty=@qty WHERE BarCode = " +
stringDataLabel, db))
                {
                    Int32 qtyDecrement = qtyValue - 1;

                    if (qtyDecrement >= 0)
                    {
                        cmd.Parameters.AddWithValue("@qty",
qtyDecrement);
                        Dispatcher.BeginInvoke((Action)(()
=> txtQty.Text = qtyDecrement.ToString()));
                        cmd.ExecuteNonQuery();
                    }
                }
                db.Close();

                Dispatcher.BeginInvoke((Action)(()         =>
DataGridRefresh()));

                Dispatcher.BeginInvoke((Action)(()         =>
txtStatus.Text = "Qty decreased"));

                }
            }
        }
    }
    catch(Exception ex)
    {
        Dispatcher.BeginInvoke((Action)(() => txtStatus.Text =
"Exception: " + ex.ToString()));
    }
}
```

All the work is performed with the barcode scan. The code actions depend on the radio button selection. If List Item is selected, the text boxes fill in the product information if found. If the New Item is selected, the user has to enter the Product Name, Price, and Qty values before scanning the barcode. Once the barcode has been scanned, the new item is stored in the database and the datagrid is refreshed. If the increment or decrement buttons are selected, the Qty for the item will be increased or decreased respectively.

14. Add the code for the ClaimedScanner_ReleaseDeviceRequested event handler method:

```
private    void    ClaimedScanner_ReleaseDeviceRequested(object?
sender, ClaimedBarcodeScanner e)
{
    Dispatcher.BeginInvoke((Action)(()    =>    txtStatus.Text    =
"Barcode being released"));
}
```

15. Add the code for the Scanner_StatusUpdated event handler method.

```
private    void    Scanner_StatusUpdated(BarcodeScanner    sender,
BarcodeScannerStatusUpdatedEventArgs args)
{
    uint bcEXStatus = args.ExtendedStatus;
    BarcodeScannerStatus bcStatus = args.Status;

    Dispatcher.BeginInvoke((Action)(()    =>    txtStatus.Text    =
"Barcode  status  update:  "  +  bcStatus.ToString()  +  "  "  +
bcEXStatus.ToString()));
}
```

16. Add POSSetup() to the MainWindow() method.

```
public MainWindow()
{
    InitializeComponent();
    POSsetup();
}
```

17. Save the project.

9.4.4   Test the Application
Now, we are ready to test the application.

1. In Visual Studio, build the application.
2. If there are any errors, correct them; and rebuild the application.

3. Make sure a USB barcode scanner is connected to the target system.
4. Start the debugger.
5. Enter some new items.

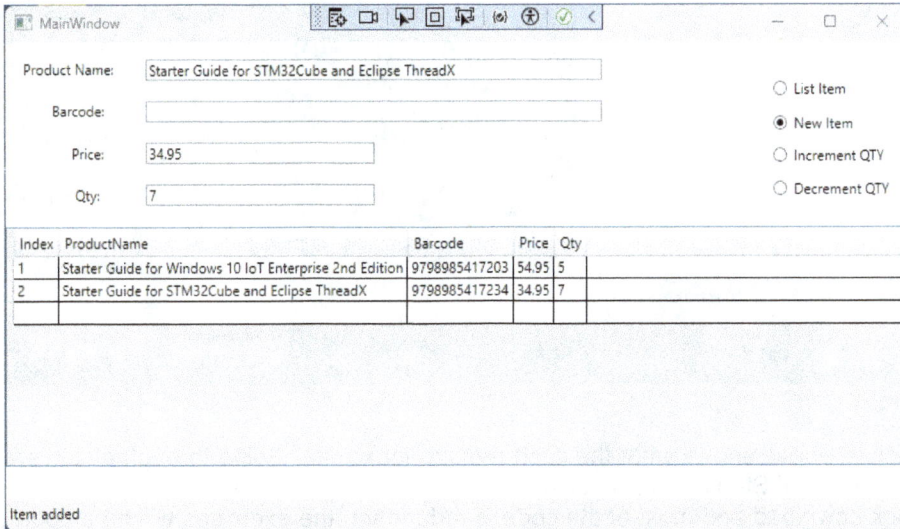

Test all controls to see that the application can scroll through the database and change quantity values for different items.

9.5 Exercise 9.4: Cash Register Review

The final exercise is to demonstrate how an application interacts with multiple POS devices. Barcode scanner, MSR, Pole Display, Receipt Printer, and cash drawer will be used with the application. The database from the previous exercise will be used to access products in inventory and update quantities.

The barcode scanner will scan items into a sales table. A total amount will be updated for all items added. Once all the items have been added, the user hits either the "Cash Payment" button or the "Credit Card Payment" button to complete the sale. For cash payment, the number pad and textbox are made available where the user enters the cash paid. Once the Enter button is pressed, the Receipt Printer prints the receipt and opens the cash drawer. For credit card payment, the textbox and Enter button are enabled. The user swipes the credit card, which displays the number in the textbox. Once the Enter button is pressed, the Receipt Printer prints the receipt. After the receipt is printed, the sale table, sale totals, and number pad are reset for the next transaction.

The exercise is only an example. Payment transactions should be handled in a safe manner. Typically, using a Pin Pad or one of the new Fin-tech solutions should be used for credit card payments. Hopefully, the example provides some ideas for your own solution.

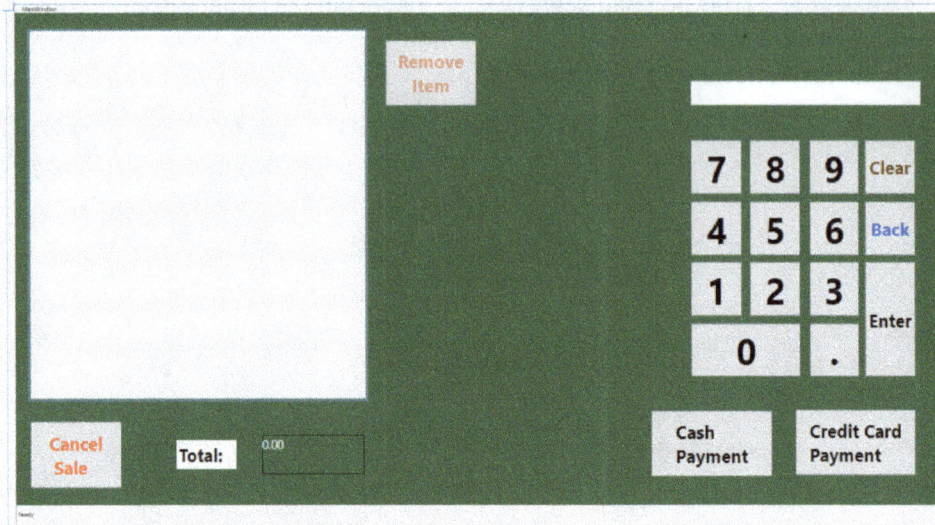

The picture above shows what the cash register looks like. Going through all the steps to create the Visual Studio project could be a little involved. Since the source code is part of the book download and most of the code is redundant, the exercises will be a code walk-through.

9.5.1   Utility.cs

After the Visual Studio project had been created, a new file resource (Utility.cs) was added to handle non-GUI methods and event handlers. With the aid of a partial class, Utiltiy.cs will be an extension of the public partial class MainWindow : Window. The POSSetup() method and all the POS device watching and event handlers are placed in this file.

```
public void POSSetup()
{
    myBarcodeDevWatcher.Added += MyBarcodeDevWatcher_Added;
    myBarcodeDevWatcher.Removed += MyBarcodeDevWatcher_Removed;
    myBarcodeDevWatcher.Start();

    myMSRDevWatcher.Added += MyMSRDevWatcher_Added;
    myMSRDevWatcher.Removed += MyMSRDevWatcher_Removed;
    myMSRDevWatcher.Start();

    myLineDisplayDevWatcher.Added                           +=
MyLineDisplayDevWatcher_Added;
    myLineDisplayDevWatcher.Removed                         +=
MyLineDisplayDevWatcher_Removed;
    myLineDisplayDevWatcher.Start();
```

```
    myPrinterDevWatcher.Added += MyPrinterDevWatcher_Added;
    myPrinterDevWatcher.Removed += MyPrinterDevWatcher_Removed;
    myPrinterDevWatcher.Start();

    myCashdrawerDevWatcher.Added                              +=
MyCashdrawerDevWatcher_Added;
    myCashdrawerDevWatcher.Removed                            +=
MyCashdrawerDevWatcher_Removed;
}
```

A data table is used to track items that are scanned in. Tkhe SalesTabletSetup() method is used to set up the columns and disable the number pad.

```
private void SalesTabletSetup()
{
    saleTable.Columns.Add("Barcode", typeof(string));
    saleTable.Columns.Add("Item", typeof(string));
    saleTable.Columns.Add("Qty", typeof(string));
    saleTable.Columns.Add("Price", typeof(string));

    //Add the columns
    for (int i = 0; i < saleTable.Columns.Count; i++)
    {
        dgSales.Columns.Add(new DataGridTextColumn()
        {
            Header = saleTable.Columns[i].ColumnName,
            Binding = new Binding { Path = new PropertyPath("["
+ i.ToString() + "]") }
        });
    }

    dgSales.ItemsSource = saleTable.DefaultView;

    btnNum9.IsEnabled = false;
    btnNum8.IsEnabled = false;
    btnNum7.IsEnabled = false;
    btnNum6.IsEnabled = false;
    btnNum5.IsEnabled = false;
    btnNum4.IsEnabled = false;
    btnNum3.IsEnabled = false;
    btnNum2.IsEnabled = false;
```

```
    btnNum1.IsEnabled = false;
    btnNum0.IsEnabled = false;
    btnNumDot.IsEnabled = false;
    btnClear.IsEnabled = false;
    btnEnter.IsEnabled = false;
    btnBack.IsEnabled = false;
    txtPaymentInput.IsEnabled = false;


}
```

The ClaimedScanner_DataReceived() method performs most of the work. When the barcode is scanned, the database is opened and a search is performed on the item. If the quantity available is not at zero, the item is added to the sale. If a duplicate item is scanned in, only the quantity is updated. If the Pole Display is available, the product name and amount will be displayed.

```
private                      async                      void
ClaimedScanner_DataReceived(ClaimedBarcodeScanner        sender,
BarcodeScannerDataReceivedEventArgs args)
{
    //Read the data from the buffer and convert to a string.
    var                    scanDataLabelReader                    =
DataReader.FromBuffer(args.Report.ScanDataLabel);
    string                  stringDataLabel                      =
scanDataLabelReader.ReadString(args.Report.ScanDataLabel.Length
);

    //Remove UPCe if capture by scanner
    if (stringDataLabel.Length > 13)
    {
        stringDataLabel = stringDataLabel.Substring(0, 13);
    }

    string dbpath = "c:\\NETPOS\\Inventory.db";

    try
    {
        using        (SqliteConnection        db        =        new
SqliteConnection($"Filename={dbpath}"))
        {

            db.Open();
```

```
        if (db.State == System.Data.ConnectionState.Open) {

            using      (SqliteCommand    cmd    =    new
SqliteCommand("SELECT* FROM Inventory WHERE BarCode = " +
stringDataLabel, db))
                {
                    using      (SqliteDataReader    reader    =
cmd.ExecuteReader())
                    {
                        reader.Read();
                        String              Item              =
reader["ProductName"].ToString();
                        String Qty = reader["Qty"].ToString();
//Get the current Qty in inventory
                        String              Price              =
reader["Price"].ToString();
                        reader.Close();

                        //First check to see if there is an item
already scanned
                        bool saleItemUpdate = false;
                        foreach( DataRow row in saleTable.Rows)
                        {
                            if
(row["Barcode"].Equals(stringDataLabel))
                            {
                                Int32        qtyUpdate        =
Int32.Parse(row["Qty"].ToString()) + 1;

                                if(!(qtyUpdate             >
Int32.Parse(Qty))){

                                    decimal extendedPriceUpdate
= decimal.Parse(row["Price"].ToString()) + decimal.Parse(Price);
                                    row["Qty"]              =
qtyUpdate.ToString();

                                    row["Price"]              =
extendedPriceUpdate.ToString();

                                    totalSale              +=
decimal.Parse(Price);
```

```
Application.Current.Dispatcher.Invoke((Action)(()                    =>
txtSaleTotal.Text = totalSale.ToString()));
                                        saleItemUpdate = true;
                                        if (myClaimedLineDisplay !=
null)
                                        {

LineDisplayTextAttribute                attribute                =
LineDisplayTextAttribute.Normal;
                                        string    displayMSG    =
Item.Substring(0,20) + " " + Price;
                                        await
myClaimedLineDisplay.DefaultWindow.TryClearTextAsync();
                                        bool    displaySuccess    =
await
myClaimedLineDisplay.DefaultWindow.TryDisplayTextAsync(displayM
SG, attribute);
                                        if (displaySuccess)
                                        {

Application.Current.Dispatcher.Invoke((Action)(()                    =>
txtStatus.Text = " Data sent sucessfull to the display"));
                                        }
                                        else
                                        {

Application.Current.Dispatcher.Invoke((Action)(()                    =>
txtStatus.Text = " Attempt to display failed"));
                                        }
                                    }
                                }
                                else
                                {

Application.Current.Dispatcher.Invoke((Action)(()                    =>
txtStatus.Text = "No more inventory"));
                                    saleItemUpdate = true;
                                }
                                break;
                            }
                        }
```

```
                        //If   the   item   hasn't   already   been
scanned, scan in the new item
                    if (!saleItemUpdate) {

                        if(Int32.Parse(Qty) >= 1)
                        {

saleTable.Rows.Add(stringDataLabel, Item, "1", Price);
                            totalSale                        +=
decimal.Parse(Price);

Application.Current.Dispatcher.Invoke((Action)(()              =>
txtSaleTotal.Text = totalSale.ToString()));


                        if(myClaimedLineDisplay != null)
                        {
                            LineDisplayTextAttribute
attribute = LineDisplayTextAttribute.Normal;
                            string     displayMSG      =
Item.Substring(0, 31) + " $" + Price;
                            await
myClaimedLineDisplay.DefaultWindow.TryClearTextAsync();
                            bool displaySuccess = await
myClaimedLineDisplay.DefaultWindow.TryDisplayTextAsync(displayM
SG, attribute);
                            if (displaySuccess)
                            {

Application.Current.Dispatcher.Invoke((Action)(()              =>
txtStatus.Text = " Data sent sucessfull to the display"));
                            }
                            else
                            {

Application.Current.Dispatcher.Invoke((Action)(()              =>
txtStatus.Text = " Attempt to display failed"));
                            }
                        }
                    }
                    else
```

```
                              {

Application.Current.Dispatcher.Invoke((Action)(()              =>
txtStatus.Text = "No inventory"));
                                    }
                      }

                    //Update the DataGrid

Application.Current.Dispatcher.Invoke((Action)(()              =>
dgSales.Items.Refresh()));
                        }
                  }
                    db.Close();
            }
          else
          {
                  Application.Current.Dispatcher.Invoke((Action)(()
=> txtStatus.Text = "Database failed to open"));
            }
        }
      }
    catch (Exception ex) {
          Application.Current.Dispatcher.Invoke((Action)(()       =>
txtStatus.Text = "Database connection error"));
      }
}
```

The other important method is the CompleteTheSale() method. This method will close out the sale by printing the receipt, opening the cash drawer, if cash is paid, updating the quantities in the inventory database, displaying a message on the Pole Display, clearing the sales table and the sales total, and disabling the numeric pad.

```
private async void CompleteTheSale()
{
    if (myClaimedPrinter != null)
    {
        if (await myClaimedPrinter.EnableAsync())
        {
```

```
            ReceiptPrintJob              job              =
myClaimedPrinter.Receipt.CreateJob();

            job.PrintLine("              My Company Store");
            job.PrintLine("         Thank you for your Business");
            job.PrintLine("");
            DateTime currentTime = DateTime.Now;
            job.PrintLine("Date: " + currentTime.ToString());
            job.PrintLine("");
            foreach (DataRow row in saleTable.Rows)
            {
                job.PrintLine(row["Item"].ToString());
                job.PrintLine(row["Barcode"].ToString()    +    "
QTY: " + row["Qty"].ToString() + " $" + row["Price"].ToString());
            }
            job.PrintLine("");
            job.PrintLine("Sales  Total:              $ " +
totalSale.ToString());
            job.PrintLine("");
            job.PrintLine("");
            job.PrintLine("");
            job.PrintLine("");
            job.PrintLine("");
            job.PrintLine("");
            job.PrintLine("");
            job.PrintLine("");
            job.CutPaper();
            if (!await job.ExecuteAsync())
            {
                Debug.WriteLine("Print failed");
            }

        }
    }
    if (!ccPayment)
    {
        if(myClaimedCD != null)
        {
            await myClaimedCD.OpenDrawerAsync();
        }
    }
```

```
    string dbpath = "c:\\NETPOS\\Inventory.db";

    try
    {
        using       (SqliteConnection       db       =       new
SqliteConnection($"Filename={dbpath}"))
        {

            db.Open();
            if (db.State == System.Data.ConnectionState.Open)
            {
                foreach(DataRow row in saleTable.Rows)
                {
                    String Qty = "0";

                    using(SqliteCommand       cmd       =       new
SqliteCommand("SELECT* FROM Inventory WHERE BarCode = " +
row["Barcode"].ToString(), db))
                    {
                        using   (SqliteDataReader   reader   =
cmd.ExecuteReader())
                        {
                            reader.Read();
                            Qty = reader["Qty"].ToString();
                            reader.Close();
                        }
                    }

                    using   (SqliteCommand   cmd2   =   new
SqliteCommand("UPDATE Inventory Set Qty=@qty WHERE BarCode = " +
row["Barcode"].ToString(), db))
                    {
                        Int32 qtyUpdateTotal = Int32.Parse(Qty)
- Int32.Parse(row["Qty"].ToString());
                        if (qtyUpdateTotal >= 0)
                        {

cmd2.Parameters.AddWithValue("@qty", qtyUpdateTotal);
                            cmd2.ExecuteNonQuery();
                        }
                    }
```

```
                }

            }
            db.Close();
        }
    }
    catch (Exception ex) {

        txtStatus.Text = ex.Message;
    }

    if (myClaimedLineDisplay != null)
    {
        LineDisplayTextAttribute              attribute         =
LineDisplayTextAttribute.Normal;
        string displayMSG = "Thank you!              Total: " +
totalSale.ToString();
        await
myClaimedLineDisplay.DefaultWindow.TryClearTextAsync();
        bool          displaySuccess          =          await
myClaimedLineDisplay.DefaultWindow.TryDisplayTextAsync(displayM
SG, attribute);
        if (displaySuccess)
        {
            Application.Current.Dispatcher.Invoke((Action)(()
=> txtStatus.Text = " Data sent sucessfull to the display"));
        }
        else
        {
            Application.Current.Dispatcher.Invoke((Action)(()
=> txtStatus.Text = " Attempt to display failed"));
        }
    }


    saleTable.Clear();
    totalSale = 0;
    txtSaleTotal.Text = "0.00";

    btnNum9.IsEnabled = false;
    btnNum8.IsEnabled = false;
    btnNum7.IsEnabled = false;
```

```
        btnNum6.IsEnabled = false;
        btnNum5.IsEnabled = false;
        btnNum4.IsEnabled = false;
        btnNum3.IsEnabled = false;
        btnNum2.IsEnabled = false;
        btnNum1.IsEnabled = false;
        btnNum0.IsEnabled = false;
        btnNumDot.IsEnabled = false;
        btnClear.IsEnabled = false;
        btnEnter.IsEnabled = false;
        btnBack.IsEnabled = false;
        txtPaymentInput.IsEnabled = false;
}
```

9.5.2 MainWindow.xaml.cs

The MainWindows.xaml.cs contains the global defines for all the POS devices and device watchers, the MainWindow() method, and all of the GUI button handlers. There are 5 methods that stand out in this file.

The first is the btnEnter_Click() method, which performs the last step in the sale. A check is performed to make sure the text box is not empty. If the credit card payment button is pressed, the call to CompleteTheSale() method is made. If the cash payment button is pressed, a comparison check on whether the total cash received/entered is greater than or equal to the sale total. If cash received is greater, the call to CompleteTheSale() method is made. What is missing here is a message to the user on how much change should be returned.

```
private void btnEnter_Click(object sender, RoutedEventArgs e)
{
    if (!(String.IsNullOrEmpty(txtPaymentInput.Text)))
    {

        if(ccPayment == true)
        {
            CompleteTheSale();
        }
        if(ccPayment == false)
        {
            decimal                    cashpaid                    =
Convert.ToDecimal(txtPaymentInput.Text.ToString());
            if (cashpaid >= totalSale)
            {
```

```
                CompleteTheSale();
        }
        else
        {
            txtStatus.Text = "Cash payment is too low to
cover sales";
        }
    }
}
txtPaymentInput.Text = "";
}
```

The btnRemoveItem_Click() method will remove a selected item on the datagrid from the sale table and update the totalsale.

```
private void btnRemoveItem_Click(object sender, RoutedEventArgs
e)
{
    DataRowView row = (DataRowView)dgSales.SelectedItem;
    if (row != null)
    {
        totalSale -= Convert.ToDecimal(row["Price"]);
        txtSaleTotal.Text = "$ " + totalSale.ToString();
        saleTable.Rows.Remove(row.Row);
        txtStatus.Text = "Ready";
    }
    else
    {
        txtStatus.Text = "Please select and item.";
    }
}
```

The btnCancelSale_Click() method clears the sale table and the sale total.

```
private void btnCancelSale_Click(object sender, RoutedEventArgs
e)
{
    totalSale = 0;
    txtSaleTotal.Text = totalSale.ToString();
    saleTable.Clear();
}
```

The btnCashPayment_Click() method enables the numeric pad so the user can enter the cash received.

```
private void btnCashPayment_Click(object sender, RoutedEventArgs
e)
{
    btnNum9.IsEnabled = true;
    btnNum8.IsEnabled = true;
    btnNum7.IsEnabled = true;
    btnNum6.IsEnabled = true;
    btnNum5.IsEnabled = true;
    btnNum4.IsEnabled = true;
    btnNum3.IsEnabled = true;
    btnNum2.IsEnabled = true;
    btnNum1.IsEnabled = true;
    btnNum0.IsEnabled = true;
    btnNumDot.IsEnabled = true;
    btnClear.IsEnabled = true;
    btnEnter.IsEnabled = true;
    btnBack.IsEnabled = true;
    txtPaymentInput.IsEnabled = true;
    ccPayment = false;
    txtPaymentInput.Text = "";
}
```

The btnCCPayment_Click() method enables only the text box and enter button. The user swipes the credit card and the account info is sent to the text box. Again, this is not the best way to handle this, but it is a demonstration only. Once the credit card is swiped, the user pushes the enter button to complete the sale.

```
private void btnCCPayment_Click(object sender, RoutedEventArgs
e)
{
    txtPaymentInput.IsEnabled = true;
    btnEnter.IsEnabled = true;
    ccPayment = true;
    txtPaymentInput.Text = "";
}
```

9.5.3 Testing the Code
Before you run the code, make sure all of the POS devices are connected and you have performed a test connection to them using one of the previous examples in the book. The

POSMGR application created in Exercise 7.6 can help view the Pos devices connected to the system. Also, make sure that you have used the previous exercise to fill up the inventory database with items.

Once you run the application, use the barcode scanner to scan items into the sale table. Click on either the cash payment or credit card payment buttons, and complete the transaction.

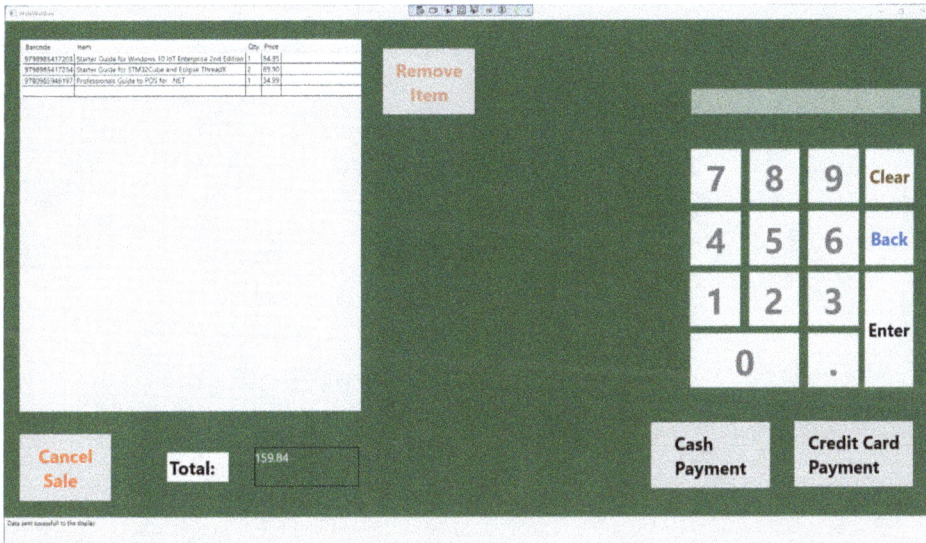

The DB Browser (SQLite) can be used to change the inventory database to see that the quantity in inventory has been reduced.

To restate the obvious, these last two exercises are just examples. More work would have to be done to make a more robust solution. To address product shipping, one improvement would be to add a weight field to the database. The InvScanner would include the support for a POS scale to record the weight of each item. For a simplistic POS device, direct COM port calls can be made rather than going through the whole POS for .NET Service Object as was done in the last chapter. The cash register application can simply total all individual item weights to generate a total shipping weight.

9.6   Summary

This final chapter demonstrated a few ideas and real-world-like applications that use multiple POS devices. The cash register application in the first book was built using POS for .NET 1.1 and Visual Studio 2005. The application ran into performance issues with all the same POS devices. The cash register application built with POS for WinRT didn't have any performance issues. POS for WinRT and .NET provides a modern approach to working with POS devices compared to the older POS for .NET and .NET Framework.

POS for .NET is still available and can be used to access the remaining 31 POS devices using POS for .NET APIs.

A Bibliography

There are a number of resources available for C# and VB.NET. This book drew from a number of books, articles, and presentations, besides the WEPOS and POS for .NET help files.

A.1   Books

Windows Embedded for Point of Service / POS for .NET, Sean D. Liming with John R. Malin, SJJ Embedded Micro Solutions, LLC. 2006, ISBN: 1-933324-54-6

Professional's Guide to POS for .NET, Sean D. Liming, Annabooks, LLC. 2011, ISBN-13: 978-0-9859461-9-7

Professional's Guide to POS for Windows Runtime, Sean D. Liming and John R. Malin, Annabooks, LLC. 2021, ISBN -13: 978-0-9911887-5-8

A.2   Microsoft Resources

Windows Embedded for Point of Service Help, Microsoft, 2005-2006

Point of Service - https://docs.microsoft.com/en-us/windows/uwp/devices-sensors/point-of-service

Windows-universal-samples - https://github.com/microsoft/Windows-universal-samples

POS for .NET v1.12 SDK Documentation, Microsoft, 2008.

Windows XP Embedded Developer Conference Presentations, Microsoft, 2006

- "Windows Embedded for Point of Services – Developing Applications"

A.3   Articles

"Implementing POS for .NET MSR for the TabletKiosk® eo a7330D", Sean D. Liming, seanliming.com, January 2010.

"What is new for POS for .NET 1.14", Sean D. Liming, John r. Malin, seanliming.com, 2014

A.4   Websites

Pipe Operations in .NET - .NET | Microsoft Learn:
https://learn.microsoft.com/en-us/dotnet/standard/io/pipe-operations

B Index

www.ingramcontent.com/pod-product-compliance
Lightning Source LLC
Chambersburg PA
CBHW081239220326
41597CB00023BA/4127